Watching Glacier's Wildlife

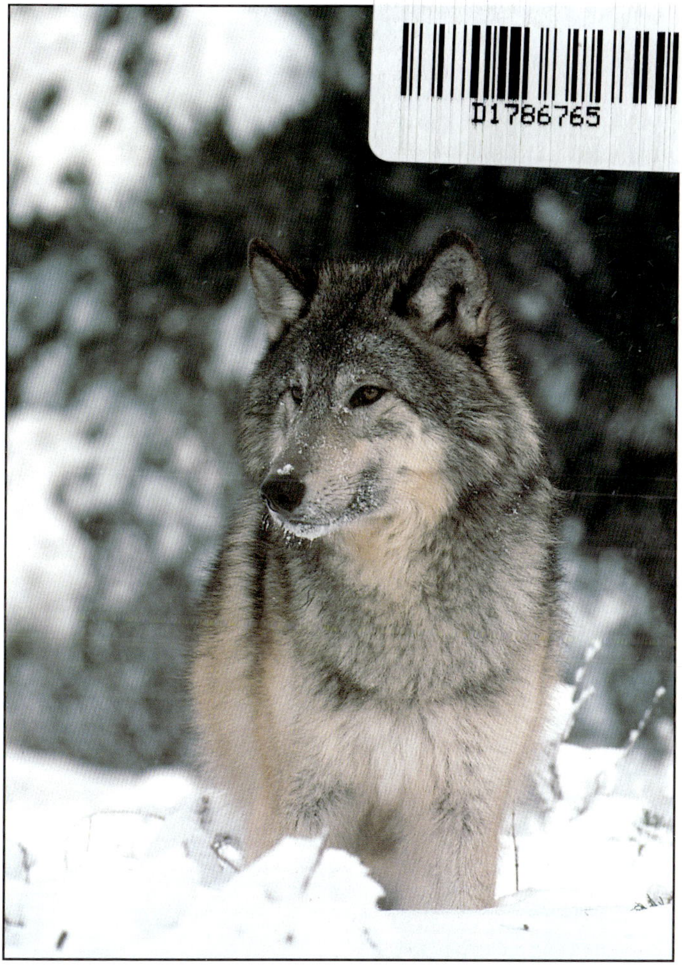

By Todd Wilkinson
Photography by Michael H. Francis

RIVERBEND
PUBLISHING

Photo identification:
    Page 1: Wolf
    Page 3: Moose
    Page 4: Mountain lion
    Page 12: Black bear cub
    Page 92: Bobcat

Copyright © 2002 Riverbend Publishing
Photographs © Michael H. Francis

Loon photographs on pages 78 and 79 © Robert W. Baldwin

Published by Riverbend Publishing, Helena, Montana.

Printed in South Korea.

1 2 3 4 5 6 7 8 9 0 SI 07 06 05 04 03 02

All rights reserved. No part of this book may be reproduced, stored, or transmitted in any form or by any means without the prior permission of the publisher, except for brief excerpts for reviews.

ISBN 1-931832-22-6

Cataloging-in-Publication data is on file at the Library of Congress.

Riverbend Publishing
P.O. Box 5833
Helena, MT 59604
1-866-787-2363
www.riverbendpublishing.com

Black-tailed Prairie Dog
(Cynomys Ludovicianus)

Introduction

These interesting rodents of Devils Tower National Monument are one of the most social wild animals of North America. They bear no similarity to dogs except their call is termed a bark like that of a dog. French Canadians named them "prairie du chien", and later English-speaking plains explorers apparently followed the French usage. Even the scientific name, Cynomys ludovicianus, perpetuates this misnomer. Cynomys is derived from the Greek word, Kynos, meaning dog.

Description

A member of the squirrel family, prairie dogs have minute ears, short tails, and muscular legs which suit them for living in tunnels. The buff-colored fur blends well with the soil of their burrows. Black hairs cover the tip of the tail and are scattered over upper parts of their bodies. The length of the adult animal varies from 11 to 13 inches (28-33 cm) and weight ranges from 2 to 3 pounds (900-1350g), with males generally heavier than the females.

Range

One of five prairie dog species, the black-tailed prairie dog once thrived on semi-arid prairie land stretching from southern Saskatchewan, Canada, to northern Mexico. However, their numbers have been drastically reduced in recent times due to extensive control and loss of habitat. Today they are found primarily in such protected areas as Devils Tower National Monument, Wind Cave National Park, Theodore Roosevelt National Park and Badlands National Park. Some privately owned lands still contain large prairie dog towns.

Population Organization

Prairie dogs are highly social and live in densely populated areas referred to as towns. Large towns are divided into wards which are separated by topographic features such as a hills, roads, streams, or tracts of trees. Wards are further divided into coteries (kot' -e-re). A typical coterie contains 1 adult male, 3-4 adult females and several yearlings and juveniles. However, coterie size can range anywhere from 2 to 39 individuals. If two adult males reside in the same coterie, one is dominant over the other. The residents of each coterie protect their territory from intruders, including prairie dogs from other coteries within the town.

Burrows

The mounds of excavated earth around prairie dog burrows serve as watch towers as well as dikes to keep out water from heavy rains. Prairie dogs repair the entrance hole by pounding wet earth into place with their noses. A burrow consists of several chambers which include a listening post (the chamber closest to the surface), a toilet, and a multichambered living area. One chamber of the living area is usually built above the bottom of the burrow and serves as an underground "lifeboat" that traps air for the prairie dog in case the burrow floods.

Young

Prairie dogs breed from late February through early April. Four to six blind and furless pups are born 35 days after breeding occurs. The mother of the young actively defends the nest burrow from all prairie dogs after the birth of the pups. Pups are nursed for 6-7 weeks, making their first appearance above ground in May and June.

Food

Prairie dogs are almost entirely vegetarian, though they will eat small insects from time to time. Tall plants are cut down both for food and to increase visibility. Preferred food plants are sometimes eliminated from the centers of long-occupied towns, leaving only a thin plant cover of short grasses and flowering plants. The main source of water for prairie dogs comes from the liquid in plants and roots of the grasses they eat. Prairie dogs neither hibernate nor store food in these latitudes. They exist over the winter principally on fat accumulated during the growing season. They are active above ground on most warm winter days where they can be seen foraging on available vegetation.

Predators

Many meat-eating animals prey upon prairie dogs for food. Predators in this area include coyote, fox, badger, mink, bobcat, weasels, owls, hawks, eagles, bullsnakes and rattlesnakes. At one time the now endangered black-footed ferret was probably an important check on prairie dog populations.

Communication

Prairie dogs communicate with each other by a variety of methods. Some of the more common include the following:

Identification kiss - The kiss is exchanged when two individuals of the same coterie meet. It is performed to show recognition and acceptance of the prairie dog in the coterie territory.

Warning bark - This bark is the most common vocalization heard at the town. It consists of a short, high-pitched bark repeated several times often accompanied by a flicking of the tail. It means the prairie dog has sensed or observed something which may be a source of danger. Individuals in the town who hear a warning bark will sit up to see what caused the alarm.

Hawk warning bark - This bark differs from the warning bark as it is a faster, more intense, of higher pitch, and of shorter duration. It will cause other prairie dogs in the immediate area to run quickly to their burrows.

Territorial call - In giving this call, a prairie dog throws its forefeet upward and points its nose straight up before coming down on all fours. The call serves as an "all-is-well" signal after danger has passed or if a prairie dog feels secure. The call can also be a warning to other prairie dogs that this territory is taken and those not in the coterie are to stay out.

Conclusion

Prairie dogs are interesting to observe as they interact with each other and their environment. **DO NOT FEED PRAIRIE DOGS!**

* Prairie dogs obtain their water solely from the plants they eat. Human food contains salt, sugar and preservatives which disrupt their nutritional requirements and upset their water balance. Prairie dogs can become quite sick and even die from a diet high in human food.

* Prairie dogs quickly become accustomed to depending on handouts for food. This dependence can cause starvation of the animal in the fall and winter when artificial feeding is not available.

* Prairie dogs can bite people who offer them food. Prairie dogs carry diseases, fleas, ticks and lice, some of which can be fatal to human beings.

* Perhaps the most important reason why feeding prairie dogs is prohibited is so that future generations who visit Devils Tower may also enjoy a <u>natural</u> and <u>healthy</u> prairie dog population.

CAUTION! Rattlesnakes and black widow spiders often use prairie dog burrows to escape the intense heat of summer days. Never reach into a prairie dog burrow, as you may be bitten by a rattlesnake, spider or prairie dog.

IT'S THE LAW!
PARK RANGERS ENFORCE THE LAW AGAINST FEEDING AND MOLESTING PRAIRIE DOGS INCLUDING FINES OF UP TO $500

Published by
THE DEVILS TOWER
NATURAL HISTORY ASSOCIATION
in cooperation with
NATIONAL PARK SERVICE
U.S. DEPARTMENT OF THE INTERIOR

Copyright, 1959, by the
Devils Tower Natural History Association

Printed by
Sand Creek Printing
Belle Fourche, SD
SCP-2380

Revised 1966.
Second revision 1990.
Third revision 1995.
Fourth revision 1997

TABLE OF CONTENTS

FOREWORD 7
WATERTON-GLACIER MAP 8
INTRODUCTION 10
WILDLIFE WATCHER'S CODE OF CONDUCT 13
TIPS FOR WATCHERS 15
WELCOME TO GLACIER 16
WILDLIFE ENCOUNTERS CHART 18

MAMMALS
Grizzly Bear 20
Black Bear 25
Gray Wolf 28
Coyote 31
Red Fox 33
Mountain Lion 34
Lynx 37
Bobcat 39
Mountain Goat 40
Bighorn Sheep 44
Moose 48
Elk (Wapiti) 50
Mule Deer 54
White-Tailed Deer 58
Bison (Buffalo) 60
Wolverine 62
Badger 63
Beaver 65
Porcupine 66
River Otter 68

BIRDS
Bald Eagle 70
Osprey 73
Golden Eagle 74
Red-Tailed Hawk 76
Common Loon 78
Harlequin Duck 80
White-Tailed Ptarmigan 82

Additional Species: A Waterton-Glacier Gallery 84
BEST WILDLIFE WATCHING HIKES 93
So You'd Like to Know More? 95

About This Guide

Glacier Wildlife: A Watcher's Guide is the first wildlife viewing guide that focuses exclusively on the mammals and birds accessible from the roadside in Glacier National Park and Waterton Lakes National Park. Not only does the book provide a catalog of individual species, it tells visitors where to look for favorite animals and provides color photographs, maps, and charts to make wildlife watching easy. The major aim of this book is to encourage readers to take a responsible approach to their roadside viewing and to avoid disturbing the wild inhabitants.

Between them, author Todd Wilkinson and photographer Michael Francis have spent more than two decades watching wildlife in the northern Rockies. While this book can't guarantee that a visitor will see a given species on each visit to the park, it's the most informative book available, and using it ensures that visitors will find and identify animals on a regular basis.

If you have comments or suggestions for future editions of this guide, please send them to: Editor, NorthWord Press, Box 1360, Minocqua WI 54548.

Acknowledgments

We extend special thanks to Amy Vanderbilt, who directs the public affairs office in Glacier. Despite working 80-hour weeks and operating with a limited staff, she managed to shepherd the manuscript through the park research, interpretation, and chief ranger's office for comments.

We recognize, too, the dedicated rangers who continue to make Glacier a special place in the national park system.

We thank Cindy Nielsen, Glacier National Park's chief naturalist, for her conscientious and invaluable review of our text.

We express our appreciation to the Glacier Natural History Association and the Glacier Institute—two exemplary organizations that have broadened public awareness about wildland issues and have built a public constituency for wildlife protection.

Finally, we offer a sentence of thanks to Glacier Superintendent Gil Lusk for having the courage to promote the idea of ecosystem management long before it became fashionable. Glacier shines as a bright light of preservation in an age when, too often, long-term ecological protection has been sacrificed for the sake of short-term political and economic gain.

Foreword

The "Shining Mountains" that constitute much of Glacier National Park's awe-inspiring wilderness experience are perhaps its best-known trademark. This fabulously unspoiled, pristine scenery, which has been referred to as the "Crown of the Continent," is also home to a vast and extremely diverse array of animal and bird life. If observed in a responsible, prudent manner, the wildlife can provide the cornerstone of an unforgettable first visit or an equally memorable return trip.

If read carefully and thoughtfully, Todd Wilkinson's wildlife watching guide can be an entry for park visitors to learn about and appreciate the wildlife—their habits and habitat—and subsequently to observe them from a safe distance.

Because Glacier is home to many powerful and potentially dangerous animals such as grizzly and black bears, mountain lions, and yes, even the deer, it is imperative that visitors understand their responsibility. Most people recognize the potential danger from a bear, but a mule deer habituated to human food handouts can also aggressively cause property damage and injury to an unsuspecting viewer.

Park managers strive to prevent wildlife from obtaining handouts that can ultimately mean the animal's demise from car wheels or worse. An adage often recited with reference to bears fits most wildlife: "A fed animal is a dead animal."

Consequently, we need your help to ensure that wild animals stay wild and well. The National Park Service could never have enough staff to catch each and every violation, accidental or intentional, but with the help of your eyes and ears, we can continue educating park visitors to keep our wildlife wild.

We humans are temporary residents and visitors to the Waterton-Glacier region. As such, we must help perpetuate the integrity of the wildlife populations that make this their permanent home.

Please take the time to familiarize yourself with Todd's code of ethics and safety recommendations. You will very likely see all the wildlife you envisioned, and you'll also know you did your part to minimize human impact on the very animals you wish to enjoy.

We can all be proud that Glacier is home to many protected mammals and birds, including gray wolves, grizzly bears, bald eagles, peregrine falcons, and more. It is possible for someone at Glacier to hear the penetrating howl of a wolf accompanied by the lonesome call of a loon. Sounds such as these stir the soul and evoke memories that can last a lifetime. If your wildlife-watching is done with care and sensitivity to the potential dangers and the needs of the wildlife, you will leave the park with unmatched memories.

H. Gilbert Lusk
Former Superintendent, Glacier National Park

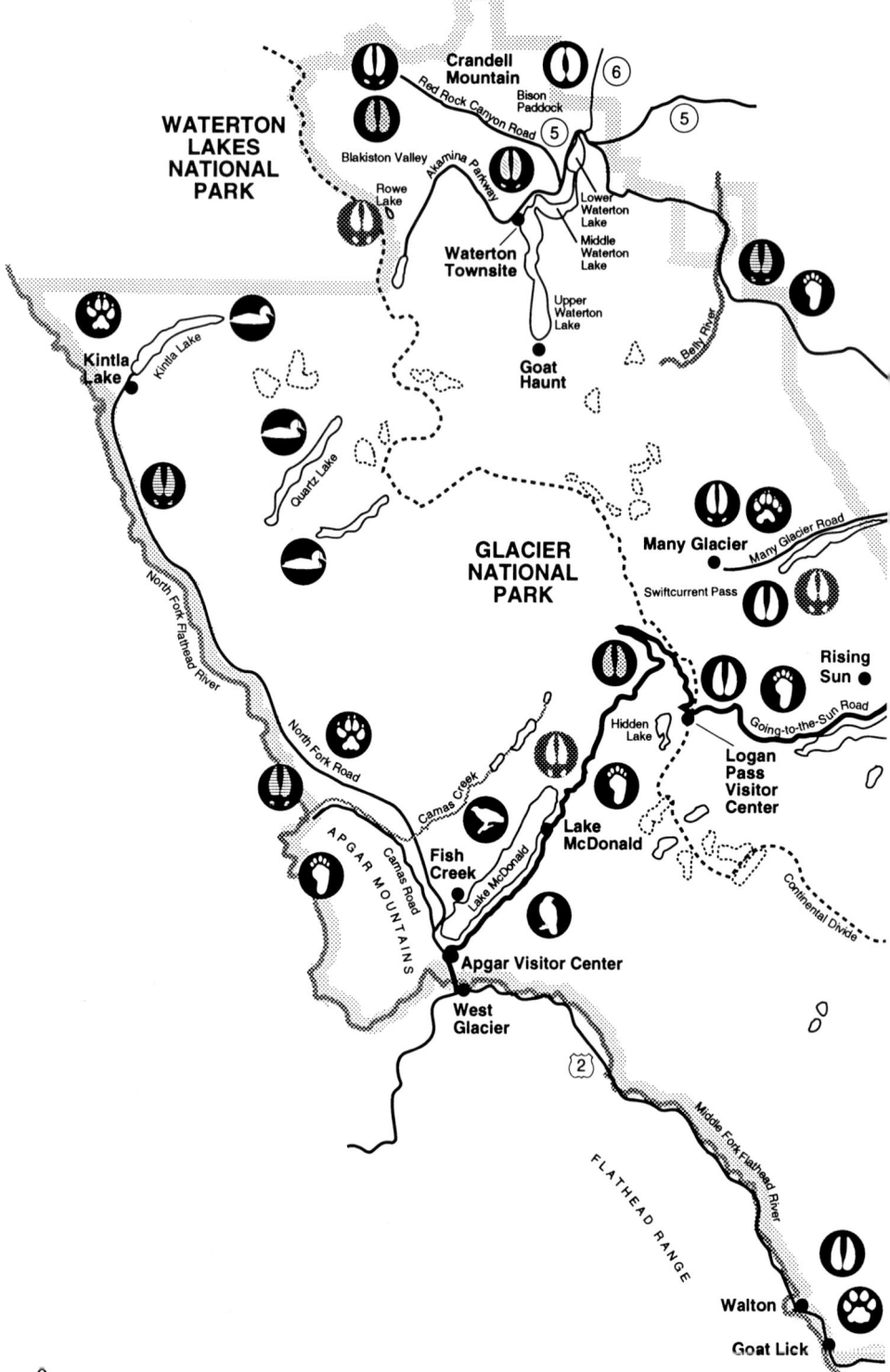

GLACIER NATIONAL PARK

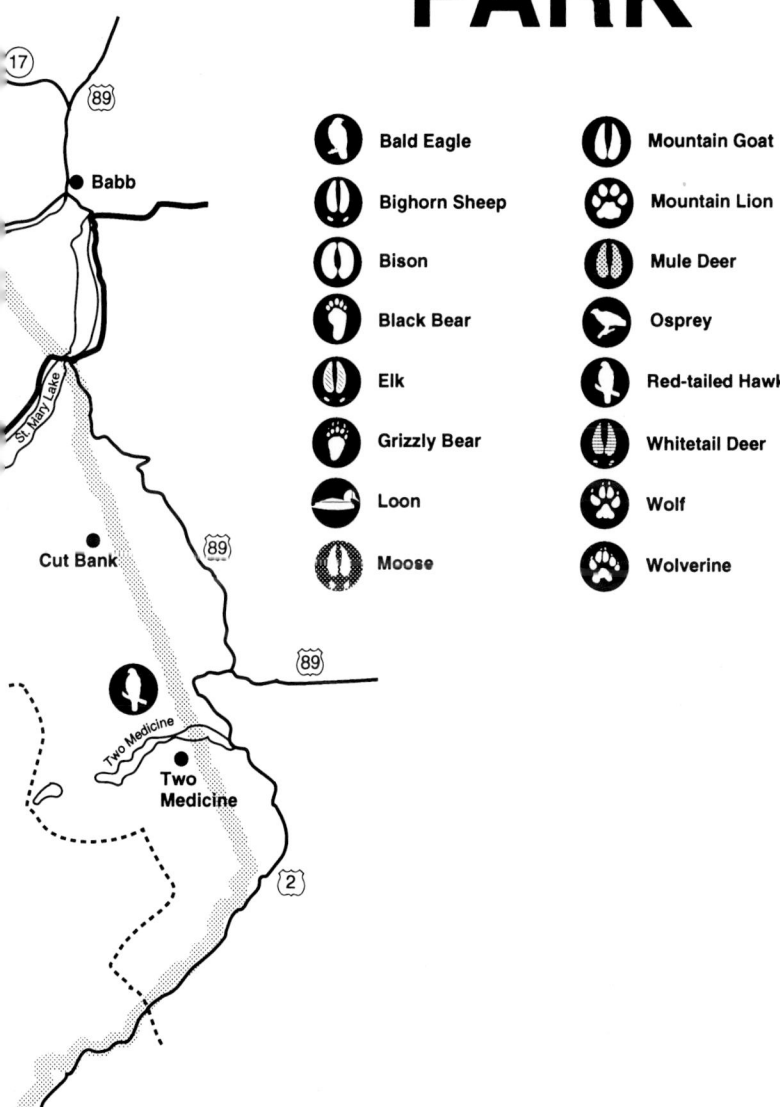

Introduction

Watching Wildlife in Glacier and Waterton
A Primeval Trek Across Two Nations

The peaks are almost too grand to be believed. The wildlife is too perfectly untamed. The waters are too absorbingly aquamarine, and yet they're gin-clear.

Few destinations in the modern world exceed human expectations, but one that does—and has for thousands of years—is Glacier National Park, situated in northwestern Montana along the United States-Canada border.

The first explorers who camped here, Native Americans hunting for wild game thousands of years ago, proclaimed this region the "land of shining mountains." Later, they passed on their admiration to trappers. Today, a necklace of 50 active glaciers remains from the post-Pleistocene era—a chain of white pearls that looms as a towering reminder of the last ice age. It's little wonder, then, that travelers navigating the mountains of Glacier National Park for the first time often compare them to the Swiss Alps, just as Theodore Roosevelt once did.

Conservationist George Bird Grinnell offered a description of Glacier National Park in *Century Magazine* that's as fitting now as it was in 1901: "No words can describe the majesty of these mountains, and even photographs seem hopelessly to dwarf and belittle the most impressive peaks," Grinnell wrote. "The fact that it is altogether unknown, the beauty and its scenery, and the opportunity it offers for mountain climbing, give the region a wonderful attraction for the lover of nature."

Still pristine and breathtaking, these geologic delights are merely a backdrop for the wild denizens that make this "Crown of the Continent" ecosystem their home—some 248 species of birds and 60 species of mammals that include federally protected species such as grizzly bears, gray wolves, and bald eagles, as well as mountain goats, bighorn sheep, and mountain lions. Some of the premier opportunities for wildlife watching in North America are found in Glacier, and this book is designed as a companion to help you find them.

Designated as a national park in 1910, Glacier does not stand alone. The 1,500-square-mile preserve abuts an older, sister sanctuary in Canada called Waterton Lakes National Park. Considered as a unit, Waterton-Glacier International Peace Park is a cornerstone of one of the largest intact ecosystems in the world. It's viewed as a model for protecting and sharing migratory wildlife across international borders.

Introduction

Of course, migratory animals do not recognize boundaries drawn on maps, but instead adhere to fixtures of topography such as river drainages, mountain ridgelines, and seasonal flying corridors.

Maintaining habitat diversity is a priority of management in both parks. In ecologic terms, the rapid rise in terrain from open plains through coniferous forests into alpine tundra makes Waterton-Glacier a rich reservoir of floral and faunal diversity. In all, five different life zones (types of terrain) are encompassed by park borders. The bilateral sanctuary is so important that each park is recognized as a Biosphere Reserve under the United Nations Educational, Scientific, and Cultural Organization's (UNESCO) Man and the Biosphere Reserve Program.

Waterton-Glacier is inexorably linked to the world around it, and one of the best illustrations is found in its complement of birdlife. "Migratory birds that spend part of the year in Glacier, for example, may depend during part of the year on habitats as far north as the Arctic Circle and as far south as Central America," writes Karen J. Schmidt in a series of scientific papers titled "Biodiversity In Glacier."

In Glacier, the west side of the Continental Divide is generally wetter, while the eastern face is more arid. What distinguishes the park in terms of animals is that it sits at the junction of the temperate pine forests common to the northern Rockies and yet brushes the belt of boreal forests native to Canada. This Canadian influence brings species to Glacier that are rare in the lower 48 states, such as the wolverine, lynx, loon, and ptarmigan.

To be sure, the bulk of Glacier's wildlife viewing opportunities occur unexpectedly—most often at the crack of dawn or through the last sunrays of dusk. Animals in the park, however, reward all humans who slow down and devote some time to the effort. The safest and by far the most pleasant experiences await the park visitor who pulls the car over and turns off the engine to hear the roar of quietness. In silence, the senses can be turned loose and allowed to run as wild as the animals themselves.

A Wildlife Watcher's Code of Conduct

Waterton-Glacier International Peace Park is listed as an official wildlife viewing site under the national "Watchable Wildlife" program, which was cooperatively initiated by government agencies and conservation organizations. As a park with this distinction, it is first and foremost the domain of wildlife, a sanctuary where humans must act sensibly and sensitively while viewing animals afield. Here are some of the basic tenets of responsible wildlife watching:

Never feed the animals

Before you consider tossing that cookie to a marmot or a grizzly bear, think again. Throwing food to wildlife is not only illegal and punishable by fines in Glacier and Waterton, but it's unethical and harmful, because it leads animals to become dependent on human handouts. This dependence can spark aggressive wildlife behavior, potentially resulting in property damage, human injury, and the unfortunate removal of the animal. In 1992, 16 bears were relocated to remote sections of Glacier because humans, using poor judgment, had fed the bears or failed to dispose of food properly. Even more tragic is the fact that several black bears were destroyed. In their quest for human food, they posed a safety hazard to park visitors. If you see people throwing scraps to animals, remind them that instead of doing the wild creature a favor, they're actually imposing a death sentence.

Maintain a safe distance

It's against the law to harass wildlife in Glacier and Waterton. Remember that both parks were set aside as wildlife preserves, and humans are guests who must respect the privacy needs of the creatures that live in the parks.

If an animal must alter its natural behavior due to your presence, you're too close. Although wildlife may appear docile, animals can become extremely dangerous if they feel threatened, particularly when a mother is trying to protect her young. Never approach bird nests, because human disturbance encourages abandonment of the nest by adult birds, and this in turn may result in predation by coyotes and other carnivores.

The rules of ethical wildlife watching should always be observed, but they take on even greater meaning during winter months, when animals are strained by the bitter cold and the lack of available food. Moose, elk, deer, and other charismatic animals are living off limited fat reserves until spring. By approaching a roadside animal and spooking it into fleeing, you may inadvertently hasten its death.

Drive with caution

The reduced speed limits in the national parks exist for a reason. Wildlife may cross the roadway at any time, and motorists should be particularly alert around dawn and dusk, when many of the larger animals are most active. If you stop to take a photograph, pull over far enough so that other cars can go around you. When traveling on narrow routes like Going-to-the-Sun Road, pull over *only* at established overlooks and parking lots. Be aware that on all national park roads, it's illegal to park in the ditch because resource damage may result.

Keep a clean camp

Many dangerous encounters between people and wildlife could have been avoided if the victims had kept a clean campsite and stored their food properly. By failing to keep your food beyond the reach of animals, you're threatening not only your own safety, but the safety of those who use the campsite after you're gone.

Park visitors are required to store their edible supplies inside their vehicle, and to make sure that all refuse is deposited in bear-proof garbage cans. If you don't have a vehicle, hang your food in a tree, using a strong plastic or cloth sack and a rope. Hang the sack at least 15 feet off the ground, over a tree limb that can't be reached by a bear.

Avoid cooking smelly foods that could attract bears and other scavengers, and never sleep in the clothes you wore while cooking. Finally, remember that all trash should be packed out of the backcountry. You should bring out everything that you brought in.

Restrain your pets

Pets—particularly barking or unruly dogs—are a liability. They affect wildlife watching because their commotion is likely to frighten the same animals you and others are trying to spot. In addition, they can cause animals such as bears, wolves, and coyotes to exhibit aggressive behavior. In Waterton-Glacier, every domestic animal must be under the owner's control and on a leash at all times. Pets are not allowed in the backcountry.

Leave antlers where you find them

Waterton-Glacier contains a variety of horned and antlered animals, but there are distinctions between them. All of the horned animals in the park, including mountain goats, bighorn sheep, and bison, are members of the bovid (cattle) family. Horns, composed primarily of keratin, remain permanently attached to an animal's skull until the end of its life.

The antler-bearing animals, including moose, elk, and mule and white-tailed deer, are part of the cervid (deer) family. These animals shed their antler branches during late winter every year, and new growths sprout in their place. It's illegal to remove any antlers found on the ground in Glacier and Waterton.

Tips for Watchers

➢ Know when to look for animals. Many wildlife species within Waterton-Glacier are nocturnal (night foraging) or crepuscular (most active at dawn and dusk). As a result, morning and evening are often the most productive times for embarking on visual safaris. At high noon, your prospects are poorest.

➢ Know the tools for safe wildlife watching. Vehicles are the safest place for humans when bears and other large mammals are near the roadside. Instead of stepping closer to roadside wildlife, consider other options.

A number of tools can enhance your view without necessitating an intrusion into the animal's space. The old standbys are a pair of binoculars or a long camera lens. Within the last decade, however, a growing number of birders and weekend naturalists have come to rely on high-powered spotting scopes that can be mounted on a car window or steadied with a portable tripod. Effective particularly when viewing grizzly bears, bald eagles, osprey, bighorn sheep, and mountain goats, spotting scopes are occasionally set up by park rangers at certain turnouts, in order to assist visitors.

Don't underestimate the value of using your vehicle as a blind. Because more than two million tourists flow through Glacier annually, the majority of large mammals have learned to accept cars when people stay inside of them.

Professional wildlife photographer Michael Francis recommends that you add a 50-300mm zoom camera lens to your equipment bag if possible. "A 300mm is a solid lens that allows visitors to get good portraits without having to move in too close," he advises. "The lens most commonly used by my colleagues, though, is a 400mm."

Visitors with cameras of any sort (including video cameras) should remember that the rules and ethics of wildlife watching apply equally to photographers.

➢ Only two of the large mammal species that were native to the area in recent times no longer inhabit Glacier—caribou and bison. However, besides the obvious complement of terrestrial animals that inhabit the park today, there are nine species of amphibians and reptiles, as well as 23 species of fish, including six that have been introduced (rainbow, Yellowstone cutthroat and brook trout, lake whitefish, kokanee salmon, and grayling). Lake trout, although native to the Saskatchewan River drainage, have also been introduced to park waters. In addition, there are at least 113 species of spiders, 364 species of beetles, 102 species of butterflies, and 16 species of Caddis flies. So far, more than 1,400 species of plants have been surveyed, and the list continues to grow. Forty-one species of plants within the park are considered rare in Montana, and 28 are endemic to Glacier—found nowhere else.

Welcome to Glacier

There are seven principal entrances to the Waterton-Glacier complex, many of them open seasonally. Once you're inside, the routes you decide to take will dictate the diversity of wildlife you'll experience. To minimize the guesswork, here are five recommended routes:

(1) Going-to-the-Sun Road provides the greatest range of species in the park and crosses five different life zones (types of terrain), but it also bears the most traffic. From Apgar Junction, the road spans 52 miles and passes along Lake McDonald at the eastern limit of Pacific rainforest species. It climbs into mountains and an old forest fire burn over Logan Pass to 6,646 feet above sea level, then drops toward the shores of St. Mary Lake and the grasslands beyond. Besides awe-inspiring scenery, visitors may encounter bears, mountain lions, moose, bald eagles, beavers, ptarmigan, white-tailed deer, mule deer, and mountain goats, among other species along this route. The Hidden Lake Nature Trail is a bonanza for wildlife watching, but stay on the boardwalk to prevent damage to the fragile alpine plant community. As a note to motorists, there are restrictions on the length of vehicles permitted on Logan Pass. However, guided tour buses and shuttles are available.

(2) The 12-mile road from Babb, Montana, to Many Glacier on the park's east side is a mecca for wildlife lovers and professional photographers. The visitor passes though grasslands and forest, with an imposing view of several peaks. From Swiftcurrent Motor Inn, there are endless locations at which you can use spotting scopes or binoculars to scan the meadows and the alpine snowfields for black and grizzly bears, coyotes, mule deer, mountain goats, bighorn sheep, and maybe even a wolverine.

(3) Take your pick. Both the Camas Road and the North Fork Road on the west side of the park are sparsely traveled and provide excellent viewing opportunities. The Camas Road, closed in winter, slants northwest from Apgar to the park entrance at Camas Creek. Whitetailed deer, bald eagles, bears, and mountain lions are sometimes spotted along the way. If you're shopping for something more rustic, take the gravel North Fork Road that ends—if you choose to drive that far—way up north at Kintla Lake, which rests at the foot of the Border Mountains. It, too, is open seasonally, but you might see bears, deer, otters and, if you're lucky, a wolf anywhere along this road. The park recommends that owners of recreational vehicles avoid driving on the inside North Fork Road, which traverses the east bank of the river.

(4) In Waterton, the Red Rock Canyon Road winds through the placid Blakiston Valley and offers prospects for spotting bighorn sheep, mountain lions, coyotes, black bears, and mule deer.

(5) Don't rule out the by-water option. The wonderful advantage of exploring for wildlife in Waterton-Glacier is that you're not limited to sitting in your car or walking down a trail. Lakes in both parks feature scenic boat trips, replete with guides who will steer you in the direction of moose, bald eagles, osprey, river otters, and other animals on your checklist. Excursions are offered from Many Glacier, Rising Sun, Waterton Lake, Two Medicine, and Lake McDonald.

Nature Trails

There are five roadside nature trails in Glacier that provide wildlife watching opportunities especially suited for families: the **Huckleberry Mountain Nature Trail** near the park entrance at Camas Creek; the **Swiftcurrent Nature Trail** near Many Glacier; the **Trail of the Cedars** north of Lake McDonald, which is accessible to visitors in wheelchairs; the **Hidden Lake Nature Trail** on Logan Pass; and the **Sun Point Nature Trail** on the shores of St. Mary Lake.

In Waterton, you can drive through the **Bison Paddock**, a fenced preserve where you'll see live bison.

Wildlife Encounters

Here's a chart that estimates the likelihood of encountering wildlife species in Glacier and Waterton. You may want to check off each species that you're able to spot in the park.

COMMON: On any given day, you should encounter one or more of this species.

IRREGULAR: By simply driving through the park without any information, you may see the species on chance encounters. However, by using the tips and suggested locations in this book, you have a good chance of seeing the animal.

RARE: While these animals have been seen in Glacier or Waterton, they are viewed so rarely that it's impossible to predict whether you'll have a roadside encounter.

MAMMALS	COMMON	IRREGULAR	RARE
Grizzly Bear		X	
Black Bear		X	
Gray Wolf			X
Coyote		X	
Red Fox			X
Mountain Lion			X
Lynx			X
Bobcat			X
Moose		X	
Elk (Wapiti)		X	
Mountain Goat	X		
Bighorn Sheep		X	
Mule Deer		X	
White-tailed Deer	X		
Bison (Waterton only)	X		
Wolverine			X
Badger			X
River Otter			X

Wildlife Encounters

MAMMALS	COMMON	IRREGULAR	RARE
Beaver			X
Porcupine			X
Hoary Marmot	X		
Colum. Ground Squirrel	X		
Snowshoe Hare		X	
Fisher			X
Pine Marten		X	
Mink		X	
Muskrat		X	
Weasel		X	
Red Squirrel	X		
Pika	X		

BIRDS	COMMON	IRREGULAR	RARE
Bald Eagle		X	
Golden Eagle		X	
Osprey			X
Peregrine Falcon			X
Common Loon		X	
Harlequin Duck			X
Ptarmigan		X	
Great Gray Owl			X
Canada Goose			X
Grouse	X		
Clark's Nutcracker	X		
Dipper	X		
Red-tailed Hawk		X	

MAMMALS

![Grizzly bear photograph]

 **The GRIZZLY BEAR**

Glacier has a global reputation as a biological island housing the largest wild predator in the lower 48 states—the grizzly bear. Glacier is better-known for its grizzlies than Yellowstone, and many travelers in the American West reroute their summer vacation in order to try to catch a glimpse of these federally protected bruins.

Everything about grizzlies makes them an anomaly in the modern world: their massive size (adults can weigh between 350 and 600 pounds); their agility (they can scale sharp inclines and outrun a horse over short distances); and the fact that they're capable of killing humans. The silhouette of a grizzly is imprinted on the minds of many as the very symbol of pristine wilderness.

Today, grizzlies occupy just one percent of their former range in the lower 48 states. Only two regions south of Canada support viable bear populations—the Greater Yellowstone Ecosystem, and the

The GRIZZLY BEAR

Crown of the Continent Ecosystem, which includes the federal Scapegoat and Bob Marshall wilderness complex, Glacier National Park, and Waterton Lakes National Park.

In 1975, grizzlies were placed on the U.S. government's list of threatened species, and were granted protection under the Endangered Species Act. Fearful that the great bear would soon become extinct in the contiguous U.S., scientists took aggressive measures to preserve bear habitat and to minimize activities that cause bear deaths. To that end, bear management in Glacier has been a success story, and the proof is evident in the number of bears that reside in the park. Rough estimates place the total population of grizzlies within the park at about 200.

The physical characteristics of a grizzly are much more pronounced than those of their smaller black bear cousins. For starters, mature grizzlies weigh at least a third more than adult black bears. Standing upright, a large adult may be up to 6 1/2 feet tall. Although the trademark of a grizzly is the silver hairs on its back that create a "grizzled" look, not all grizzlies are silvered. The hue of a bear's coat can vary widely from animal to animal. Some are cinnamon-colored, others light brown or blonde, and still others may be black.

You can generally distinguish an adult grizzly by its concave face, which resembles a dish. Look also for a shoulder hump of well-defined muscle behind the neck that's used for digging roots, insects, and small rodents out of their burrows. Besides these natural snacks, grizzlies grub for army cutworm moths, and may eat whitebark pine nuts, glacier lilies, and other tubers.

Grizzlies are proficient swimmers, and they *can* climb trees. Even when a bear is not in sight, a mindful wildlife watcher will be alert for bear presence. There are three relatively obvious signs that a bear has been in the area.

The first indicator is bear tracks in the mud. A typical adult's track is at least eight inches long, and reflects the bear's classification as a "plantigrade," or flat-footed walker. The front and rear paw tracks look almost like a human's footprint, revealing five toes plus the indentations from their claws.

Another sure sign of bears is their scat, which is tubular and usually dark-colored, reflecting the animal's diet. During summer and autumn, when bears are gorging themselves on berries and plants, the scat may be diarrhettic.

The third indication of bear presence is scratch marks on trees, made by the bear's long claws.

The wide-ranging grizzly bears are omnivores, meaning that they eat both meat and plants. They are the second-largest omnivore in

North America, dwarfed only by the polar bear. When taxonomists first explored grizzly country during the 19th century, the bear's brawny size and aggressiveness inspired its scientific name, *Ursus arctos horribilis*. Labeling the bear "horrible," however, is a rather unscientific exaggeration.

Grizzlies are long-lived animals compared to other species in the park. A healthy old-timer may be 25 or 30 years old. Most bears do not breed until they're at least five years old. After females become impregnated in early summer, development of the embryo is delayed until autumn, about the time the females are ready to den. During hibernation, bears slumber in a state of near-suspended animation, in which the metabolic rate and heartbeat slow down. Nonetheless, bears can awaken.

During winter dormancy, one to four cubs are born, though the usual number is two. At the time of birth, the cubs may weigh a pound or less apiece, but they rapidly grow to between 10 and 20 pounds by late April or early May, when they emerge from the den. Cubs usually stay at their mother's side for two summers, then set out to establish their own territories. It should be noted that females do not breed while raising their cubs, so three years pass before they once again carry young. The slow reproduction cycle is one of the reasons why protecting sow grizzlies is crucial to maintaining a viable bear population.

The grizzlies of Glacier can strike fear into those who plan on taking an overnight trip through the backcountry. While confrontations between people and bears are extremely rare, there have been incidents that will forever remind hikers to exercise caution. Perhaps the most famous bear attacks in the contiguous U.S. are those that occurred in Glacier on August 13, 1967. The attacks and the subsequent deaths of two campers were made part of park history in Jack Olsen's book, *Night of the Grizzlies*—required reading for anyone who doubts the need for taking precautions in bear country. Another useful book, considered the bible on human-bear confrontations, is Stephen Herrero's *Bear Attacks: Their Causes and Avoidance*.

Whether you decide to camp in a developed area along the road or at a site miles from anywhere, there are common sense rules for reducing the odds of having a too-close bear encounter:

➤ Avoid cooking and storing smelly foods.
➤ Don't sleep with food in your tent.
➤ Change into clothing other than the garments you wore while cooking.
➤ Keep a clean camp, and stow your edibles in the trunk of your vehicle.

The GRIZZLY BEAR

Generally, roadside wildlife watchers can avoid most risks by remaining in their car with the windows rolled up when a bear is in the vicinity. The topography of Glacier, while demanding for backpackers, works to the advantage of bear watchers along the road. With binoculars, a spotting scope, or a long camera lens, a tremendous amount of bear terrain can be surveyed from a roadside pull-over, and some spots produce regular sightings.

The best contribution a visitor can make to bear conservation is taking action to prevent an encounter from occurring, and reminding others that a bear habituated to people and human food may have to be killed or removed from the ecosystem. Because Glacier's grizzly population numbers about 200, each individual bruin is a significant member of the overall population.

Where to look for grizzly bears

Never, under any circumstances, should you approach or feed a grizzly. These bears are dangerous, and feeding them does them more harm than good. Besides, feeding bears is against the law.

Look for grizzlies at these locations:

➤ At Many Glacier, park rangers sometimes set up a spotting scope in the Swiftcurrent Motor Inn parking lot to scan the hillsides for grizzlies, especially above tree line and on both the northern and southern slopes that flank either side of the valley.

➤ In spring and early summer, grizzlies can be seen off of Going-to-the-Sun Road, wandering across snowfields and feeding on vegetation, particularly if you look toward Virginia Falls on the east side of Logan Pass.

➤ Along the Camas Road just inside the park entrance at Camas Creek, grizzlies are sometimes seen in the summer as they converge to eat ripening huckleberries.

➤ Grizzlies may be seen on Swiftcurrent Pass, en route to Granite Park Chalet.

➤ In Waterton, grizzlies are occasionally sighted on the slopes rising over Akamina Pass, near the end of Akamina Parkway.

 ## The BLACK BEAR

Although grizzlies grab the majority of newspaper headlines, the most common bruin in Glacier National Park is a species native to most regions of the United States—the black bear. Wildlife biologists estimate the number of black bears in the park at 500—about 2 1/2 times the size of the local grizzly population.

Despite the black bear's prevalence, most bear research over the last three decades has focused on grizzlies, due primarily to their imperiled status. This does not mean that black bears are unimportant in the ecosystem. As large predators and scavengers, they occupy an important niche in the food chain. Perhaps because of their numbers, black bears have caused more injuries to humans over the years than grizzlies.

By the very meaning of their scientific identification, *Ursus americanus*, black bears hold title to the name "American bear." In Glacier, they're easy to identify and fairly well dispersed. Perhaps the greatest error people commit in scanning the roadside, however, is assuming that all black bears are black. Depending on the genetic makeup of their parents, offspring may be brown, blonde, tawny, reddish, chocolate, or jet-black. Visitors routinely mistake lighter-hued black bears for grizzlies.

Generally, the facial characteristics and physical outline of a black bear are markedly different than a grizzly's. The black bear's muzzle is longer, culminating in a "Roman nose," and its head in profile is conical-shaped. The grizzly has a flatter visage that's circular, like a dish.

The smaller size of the black bear also sets it apart from grizzlies. Adult boars weigh between 200 and 400 pounds; sows weigh 150 to 300 pounds. Noticeably absent from a black bear's back is the classic shoulder hump that adorns a grizzly's upper spine.

The amateur naturalist can watch for signs indicating that a bear has been using an area. The most obvious sign is tracks. Because the bear is a "plantigrade," or flat-footed walker, the outline of its paw resembles a human foot. The impression of five toes is visible in mud, but the pawprint is wider than a human's footprint, and it usually contains claw marks in front of the toes. The tracks are generally no more than seven inches long.

Another sure indication of bear presence is scat (fecal droppings). Tubular-shaped (similar to a human's) and dark-colored, scat reflects the bear's diet. When the animal is feeding on carrion, for example, scat is firm; the texture is diarrhettic when the animal nourishes itself on berries later in the summer.

Black bears are not nearly as likely to prey on larger mammals as grizzlies. Instead, they eke out a living as exceptional scavengers of fish, small rodents, and animal carcasses, and by grubbing for insects and plants.

In one respect, black bears have a profound edge over grizzlies: They're much better at climbing trees. Their claws are shorter and sharper than a grizzly's, allowing black bears and their cubs to use the abundant pine trees in the park to escape danger. When provoked, black bears have been known to follow humans up a tree, so it's important to do everything possible to avoid a confrontation—particularly with sows and their cubs. The same rules of safety associated with watching grizzlies apply to black bears. Females are aggressively protective of their cubs, so it's wise to announce your presence on a backcountry trail by making loud noises.

Nature has bestowed bears with a unique means of reproducing.

The BLACK BEAR

Although sows and boars breed in early summer, development of the embryo is delayed until autumn, when the expectant mother goes into her den. During the long period of winter dormancy, between one and four (but usually two) fetuses grow in the mother's womb. Cubs are born toward the end of winter, and they will grow in the den from a weight of about 12 ounces each to more than ten pounds by the time they see their first daylight. Cubs usually remain with their mother until their second summer, then begin to forage and den on their own. Like grizzlies, black bears are long-lived and have reached documented ages of 20 years.

Black bears are hunted outside the park in Montana, but Glacier's prohibition on hunting affords them year-round refuge. Still, poaching of bears to satisfy a global market for gall bladders, claws, and trophy heads has seriously imperiled several populations of black bears across the country. At both Glacier and Waterton, citizens play a major role in the bear's protection, and any suspicious activity should be reported at once to park rangers.

Where to look for black bears

Never, under any circumstances, approach or feed black bears. These animals are dangerous, and feeding them does the animals more harm than good. Besides, feeding wildlife is against the law.

To see black bears, visit the following locations in the early morning, or in the evening hours just before sunset.

➢ Roadside sightings have been reported at Lower McDonald Creek, between Apgar Visitor Center and West Glacier, and along the Camas Road in the vicinity of Fish Creek.

➢ At the Swiftcurrent Motor Inn, park rangers sometimes set up a spotting scope in the early evening hours to help visitors identify both black bears and grizzlies on slopes visible from the parking lot.

➢ Along Going-to-the-Sun Road just north of Lake McDonald and on the east side of Logan Pass, black bears are sometimes seen in the vicinity of Rising Sun.

➢ In Waterton, try looking for black bears along Chief Mountain International Highway (Highway 17) south of Belly River just across the U.S. border; and along the Red Rock Canyon Road before and after Crandell Mountain.

The GRAY WOLF

Gray wolves (*Canis lupus*) have returned to Glacier after a conspicuous absence lasting several decades. Today, Glacier is the only national park in the western U.S. able to make this extraordinary claim. A new generation of wildlife watchers now has the opportunity to learn that wild wolves—contrary to mythology and misguided human attitudes—actually play an important role in bringing balance to park ecosystems.

Public opinion about these primal howlers has changed radically since the early part of the 20th century, when many predators (including wolves, bears, eagles, and mountain lions) were killed in order to eliminate an alleged threat to other wildlife and domestic livestock. Over 100,000 wolves were bountied and destroyed in Montana between 1883 and 1942. As a result of the systematic persecution, practically no wolves inhabited the park until recently.

In 1986, progenitors of a pack appropriately known as "The Magic Pack" crossed into Glacier from the Canadian province of British Columbia. Within a year, four wolf packs were known to range across areas within the Crown of the Continent ecosystem, and today the number of wolves inhabiting northern Montana is growing. Already, the management of wolves in Glacier has been hailed as a model illustrating ways in which humans can cope with wolves in Yellowstone National Park, should the animals continue to migrate southward down the spine of the Rockies.

Glacier's modern rendezvous with wolves did not happen by accident. The park's isolation, its ready prey base of deer, elk, and moose, and its commitment to protecting native wildlife made it an attractive destination for roving wolves.

For cultures that have age-old ties to the park, the return of wolves was long overdue. Native Americans evolved for thousands of years in close proximity to wolves, and the elusive predator is a central figure in Indian folklore. In the oral traditions of the Blackfeet tribe, which has a reservation bordering Glacier to the east, wolves are revered as purveyors of "good medicine." The Blackfeet express their love for wolves with a simple aphorism: "The gun that shoots at a wolf or coyote will never again shoot straight."

While the fears expressed by ranchers that wolves will kill some of their livestock are warranted, stories that portray the gray wolf as a threat to humans and wildlife populations in the park are unfounded. In modern times, not one documented case has occurred in North America in which a healthy, wild wolf attacked or killed a human. Further, studies suggest that wolves will not have a significant long-term impact on resident populations of deer and elk. In fact, scientists say that preying wolves may actually bring equilibrium to ungulate populations. Pack sizes range from roughly six to twelve animals, and adult wolves consume between five and twelve pounds of meat daily, meaning that a single elk could sustain a pack for a week.

The future of wolves in the park depends on the successful production of pups, which are fathered by an alpha male—the male pack leader—and his mate, the alpha female. They are generally the only pack members that breed, doing so in February or March. They produce a litter of four to seven pups in a den two months later. Wolves

do not hibernate, and use the den only for raising their young.

According to biologists, a number of years may pass before wolves are viewed routinely from the roadside or heard howling as hikers set up camp in the backcountry. Wolves remain scarce, and they are usually quite wary of humans. People who venture into wolf country for the first time often think they've spotted a wolf, when in actuality they're watching a coyote. The two canids are similar, though a typical adult wolf weighs roughly 90 pounds and is three times heavier than a coyote. More than 2 1/2 feet tall at the shoulder, individual wolves vary in color from gray to black, brown, and even white. Fifty percent of the wolves in the Camas Pack are black; the others are gray-hued. Compared to coyotes, wolves have a wider face, a less-tubular snout, and shorter ears that are less overtly triangular.

Unless you're lucky, the closest encounter you're likely to have with a wolf will involve studying its fresh tracks, which resemble a domestic dog's. You'll see the impressions of four toes, a paw pad, and claw marks. The imprint should measure between 4 1/2 and 5 inches in diameter.

A wolf begins reaching old age in its seventh year, and normally will not live past the age of ten. The oldest wolf on record lived for 16 years. Park visitors can aid researchers in the management of wolves by reporting all sightings as soon as they occur, including the location and the time of day.

Where to look for wolves

➤ Wolves may be sighted along the North Fork of the Flathead River between the road to Bowman Lake and Kintla Lake. However, because wolves are wide-ranging animals and the number of the packs is likely to increase, wolves may turn up almost anywhere in the park. They are most active at dawn and dusk.

➤ Be aware of the possibility that you may spot wolves as far south as the Fish Creek area along the Camas Road.

# The COYOTE

The singing banter of coyotes is a melodic treat that every human should hear at least once. A fascinating exchange of information, it colors our nights with the same language that spoke to *Homo sapiens* thousands of years ago as they moved across the open spaces of North America. More importantly, coyote yips indicate the presence of one of the most socially interactive species in Glacier National Park.

Coyotes (*Canis latrans*) have earned a place in Native American legends as the famed "trickster," but this reputation is not built on deception. Rather, it acknowledges an intelligent ceature that today is found from the Atlantic to the Pacific, from serene sanctuaries like national parks to the urban centers of Los Angeles and New York City. Coyotes are survivors that have adapted to most forms of adversity thrown their way, including a federal campaign waged earlier in the 20th century to eliminate as many coyotes and other predators as possible.

Until the recent return of wolves, coyotes were the top predatory dogs in Glacier. As the number of wolves increases, the coyote population will most likely decline, because the larger cousin will usurp hunting territories. Rarely, biologists say, will a wolf pack tolerate coyotes. A given geographic area with an abundance of prey supports eight to ten times as many coyotes as wolves, but the two species are competitors.

Wildlife watchers routinely confuse coyotes with wolves, but you can distinguish between the species with a little knowledge of their features. Coyotes weigh about a third as much as wolves—30 pounds compared to 90. Coyotes are leaner, and they sport fluffier tails. Also, coyotes sprint with their tails pointed downward, while wolves run with their tails held parallel to the ground. As the seasons progress, coyotes undergo color changes as a means of camouflage. In spring and summer their coats are brownish, but the fur turns gray as winter advances.

Tracks are a primary indicator of coyote presence. Similar to a wolf's track but smaller, a coyote track is 2 1/2 inches long and highlighted by four toe and claw imprints.

Coyotes are remarkably clever. Individuals have been known to accompany badgers as the smaller animals dig into the burrows of ground squirrels. While the badger plows through the front entrance of a squirrel's den, the coyote waits with open jaw at a secondary escape hatch. It's often rewarded with a "squirrel snack."

Coyotes are rovers known to trot dozens of miles in small groups looking for food—be it snowshoe hares, rodents, grasshoppers, carrion from deer, elk, moose and bighorn sheep, or the eggs and hatchlings of ground-nesting birds. In spring, coyote females dig dens in which they give birth to pups after late-winter mating. The male partner may remain a female's breeding companion for life.

Coyotes within national parks are unique, because they're not subject to trapping by federal agents who kill predators as a means of protecting livestock. The trapping is a tradition that goes back almost a century. At one time, coyotes inside Glacier were destroyed by bounty hunters, but all animals in the park are now protected.

Where to look for coyotes

Because they're adaptable, coyotes can easily become habituated to people and human food. Such scenarios can lead to dangerous confrontations in which coyotes aggressively approach people for handouts. Resist the temptation to throw coyotes your table scraps. Not only is this illegal and punishable by a fine, but it can lead to human injury and the necessary destruction of the animal by rangers.

➤ Coyotes are often sighted in meadows or heard yipping just south of St. Mary Junction along Highway 89 at Two Dog Flats.

➤ The road leading to Many Glacier, especially between the west end of Sherburne Lake and Swiftcurrent Motor Inn, yields frequent sightings.

➤ In Waterton, listen for coyotes later in the evening along the Red Rock Canyon Road past Crandell Mountain. Also, a resident coyote group roams the meadows at the northern end of Upper Waterton Lake, near the intersection of Highways 5, 6, and 17.

The RED FOX

The smallest wild canine in Glacier is also the least numerous and most timid. The red fox (*Vulpes vulpes*) is distributed across four-fifths of the North American continent, but its presence within the national park is limited.

Active both day and night, the red fox maintains a low profile in middle-elevation meadows tucked well away from the roads criss-crossing the park.

The animal's uncanny ability to outwit pursuers, including dogs, hunters, and trappers, is evident in Glacier's fox population. Here, it's a sly and cunning scavenger that has acquired a number of special behavioral adaptations. One adaptation is its high proficiency when mousing and stalking rabbits. Another is its ability to scavenge on carcasses left behind by other predators. As wolves re-establish a permanent foothold in the park, the number of added deer and elk carcasses could feed foxes and bolster their numbers. Researchers say that, while coyotes may be driven out by territorial wolves, the red fox should suffer little, because its survival does not depend on the same prey.

Though all are canids, you're not likely to mistake a red fox for a coyote or a gray wolf. The fox has a slender frame with a classic, billowy tail at least a third as long as its body. Although the majority of foxes have furry coats dominated by an orange-red hue, color variations do occur. Foxes may be black, silver, or reddish-brown.

Where to look for red fox

➢ Red foxes are sometimes seen on autumn afternoons, chomping on grasshoppers or mousing. Look for them in the meadows falling away from the Going-to-the-Sun Road.

➢ Along Highway 89 south of St. Mary and near Divide Creek, red foxes are occasionally sighted in the early morning or evening. Also, look for them north of St. Mary to the junction of Many Glacier.

 **The MOUNTAIN LION**

During the 19th century, explorers of the Rocky Mountain West wrote in their journals describing quiet nights around the campfire that were interrupted by the hair-raising cries of felines echoing through the forest. While some of these wilderness recollections may

The MOUNTAIN LION

have been embellished, the aura of mystery surrounding the mountain lion and its cry remains intact to this day.

As the largest of the wildcats on the continent, the species (*Felis concolor*, or "cat of one color") goes by many colloquial names: cougar, catamount, puma, panther, and, of course, mountain lion. These big cats are present in Glacier, but their stealthfulness makes them difficult to track. It's nearly impossible to predict when they'll appear before human travelers, and mountain lion sightings are so rare that visitors seldom capture a glimpse of the buff-colored carnivores.

In the United States, an estimated 16,000 mountain lions range from the Pacific Northwest to the lower Appalachians. Mountain lions have survived in Glacier despite assaults by bounty hunters who tried to eradicate all predators from western public lands earlier in this century. The animals survived due to their own elusiveness and the fact that the park offers an abundance of the things they eat—deer, young elk, small mammals, even family pets upon occasion. Today, the population appears to be rebounding, and the park is launching a mountain lion research project in Glacier's North Fork drainage to help biologists learn more about these big cats.

Just looking at the tools nature has given these animals helps an observer understand why they're able to exist in some of the harshest terrain. Muscular yet sleek, adult mountain lions weigh between 90 and 150 pounds, with males weighing more than females. They move virtually unnoticed across the bouldered slopes of Glacier, and their deceptive speed makes them a formidable threat to prey. Equipped with long, retractable claws, they can climb trees as easily as they scale vertical cliffs.

The average adult female and its young will subsist on roughly one mule deer or white-tailed deer per week. Populations of lions correspond to changes in the number of available deer and elk. If deer are not present in sufficient numbers to satisfy a mountain lion's hunger, the animal will turn to smaller mammals, including beavers, hares, birds, even porcupines and domestic livestock. Biologists report that cats often lie in wait along game trails, ambushing their prey by leaping on the backs of unsuspecting victims from more than a dozen feet away.

Generally, young lions leave their mothers one to two years after birth and establish a solitary existence, setting out on nightly hunting excursions that may cover 25 miles. Despite estimations that Glacier sits in the middle of one of the highest mountain lion concentrations in Montana, experts say they exist in low densities here compared to other large predators, and they establish territories that may extend 200 miles. Humans are considered the lion's only natural enemy, but during harsh winters when food is in short supply, they have been known to battle with coyotes over a carcass.

Mountain lions have not traditionally posed a threat to human safety, but in recent years several incidents between people and lions have forced the park to issue warnings. A 1986 study found that only three humans were attacked in the U.S. between 1750 and 1986. "Statistically, mountain lion attacks are much less likely than domestic dog attacks, or even shark attacks," writes Kathy Dimont in a publication entitled *Big Cats*. "In fact, more people are hurt by deer than by mountain lions."

Nevertheless, in 1990 and 1992, young boys were attacked and injured by lions on the shores of Lake McDonald. Rangers say park visitors should be especially wary of lions that show no fear of humans and walk into developed areas. The recommendations for confronting a curious lion are slightly different than those for bears. Never run from a lion you meet in a campground or on a trail. Try to act as aggressively as possible. Do not crouch or "play dead." Lions can sometimes be repelled by rocks or sticks thrown to impede their advance.

Mountain lions are active year-round. In areas like Glacier, where they're not hunted, they may stalk prey in the open during daylight hours, though research shows they're busiest under the cover of darkness.

Signs of a mountain lion include copious scat ranging from dark spheres to pellets, often containing traces of whatever the lion ingested, such as deer hair or bones. The tracks, measuring between three and four inches in diameter, are wider than they are long, with the imprint of four toes like those of a housecat. The cats move in front paw/back paws pairs, so the imprints are staggered, with front pawprints overlapping those of the hind paw. Remember, however, that your chances of seeing a mountain lion are extremely remote.

Where to look for mountain lions

Never, under any circumstances, should you approach or feed mountain lions. These animals can be aggressive and dangerous.

➢ Mountains lions have been seen (though rarely) at Goat Lick on the southern border of the park. During one recent sighting, a wildlife photographer watched a lion attack a mountain goat.

➢ On some evenings, rangers assemble a spotting scope in the parking lot of the Swiftcurrent Motor Inn. Using the scope, visitors have seen both mountain lions and bears on the rocky slopes rising above the Many Glacier area.

➣ On the shores of Lake McDonald, north of Lake McDonald Lodge, and near the Trail of the Cedars, mountain lion sightings have been on the rise in recent years.

➣ In Waterton, lions have been known to inhabit the craggy slopes that rise above Red Rock Canyon Road northwest of Crandell Mountain.

The LYNX

The lynx is a denizen of northern pine forests. Its presence in Glacier is proof of Canada's influence on the park. Lynx are wide-ranging carnivores whose scientific name (*Lynx canadensis*) means "Canada cat." Scientific evidence suggests that lynx routinely pass back and forth across the international border as they engage in solitary hunting sojourns between Glacier and Waterton Lakes.

Lynx bear an obvious resemblance in size and color to bobcats, though the two roam across somewhat different ranges. If you're lucky enough to see one of the two creatures, chances are it will be a lynx. The surest means of telling them apart is to find the trademark tufts of black hair that stick out prominently from the tips of the lynx's ears. Lynx also have an unmistakable tuft of scruffy, black-striped fur beneath their chins, and their generally grayish, mottled coat turns brownish during the summer. Both cats have stubby tails.

Highly adept at surviving long winters, the lynx has broad paws that serve as snowshoes for navigating the deep snows that blanket its native woodlands for six months at a stretch. But the feathery feet serve another vital purpose. They allow the animal to trek noiselessly across snow in pursuit of its favorite prey—snowshoe hares and other small mammals. The four-toed track of an adult lynx is roughly four inches in diameter and about twice as large as a bobcat's. Both resemble smaller versions of a mountain lion print.

A furtive prowler that can easily attain a body weight of 20 pounds or more, the lynx is chiefly a nocturnal hunter whose stalking habits parallel those of the mountain lion and bobcat. By day, a lynx will rest beneath a rock outcropping or within the tangled roots of a downed tree. From this camouflaged perch, it ambushes hares, upland fowl, and small rodents. Because it's not particularly swift, the lynx relies on the element of surprise when taking prey. In both Glacier and Waterton Lakes, local lynx populations correspond directly with the abundance of snowshoe hares; when hare numbers are highest, lynx and lynx sightings increase.

Legal trapping seasons in nearby states and Canadian provinces have been one cause for concern about the lynx in recent years. The animal is protected in parks, but its population can be affected when roving individuals are killed and harvested for their furs on lands outside park boundaries. Further, while the states of Idaho and Wyoming have marked the lynx for protection because it's so rare, habitat destruction continues to place the long-term survival of lynx in question, particularly in the contiguous U.S.

While lynx are more common than bobcats in Glacier, sightings are of great interest to park officials. If you see a lynx, record the time and place and report it to rangers. The information you provide will enhance protection of the species.

Where to look for lynx

Never approach or try to feed lynx.

➤ Lynx are careful to stay beyond the view of humans and other large predators during daylight hours. Their nocturnal habits further reduce the likelihood of spotting them from the road. You have the best chance of seeing them by waiting at dusk and dawn in areas frequented by snowshoe hares.

The BOBCAT

The BOBCAT

Glacier National Park is located along the northern edge of bobcat country, and the park is considered more of a sanctuary for the bobcat's cousin, the lynx. However, bobcats (*Lynx rufus*) have inhabited open areas of the park where there are few lynx. Like mountain lions, bobcats thrive on the forest's edge, where there is an abundance of rocky, south-facing slopes and meadows. The climate in these places is warmer, and the craggy setting provides shelter as well as hiding places bobcats use to surprise their favorite prey—rabbits, grouse, small rodents, even deer fawns and elk calves if given an opportunity.

As many as 1.5 million bobcats inhabit the U.S. While lynx are creatures of the boreal forest, bobcats find a niche in more temperate woodlands and sub-alpine meadows, within a region that extends from southern Canada into Mexico. The bobcat is the most common wildcat in North America.

You can identify a bobcat by a thin sliver of black hair that extends from the top of its ears, but the tail gives the animal its name. Knobby and "bobbed," its tail is no more than five inches long, and several black rings run across the brownish fur.

The cat's other distinguishing marks are the leopard-like spots on its sides and legs. The outer flanks of its underbelly are mottled, while the center of its chest is creamy white.

The brown coloration that dominates the top of a bobcat's coat changes to a grayish tone as winter approaches. Because their paws are smaller and narrower, bobcats are less mobile in deep snow than lynx.

Bobcats were not spared from the persecution heaped on predators in the early decades of this century by bounty hunters, who were hired to eliminate all threats to livestock. The bobcat population was so reduced until recently that localized extinctions were recorded, and only today is the species experiencing a recovery—albeit a slow one. The bobcat is still hunted in most Western states, including public lands around Glacier Park.

Where to look for bobcats

Never approach or feed bobcats.

➤ These animals are reclusive and nocturnal, which means there's little chance of seeing them during the middle of the day. Rare sightings of bobcats have occurred along lakes that spur off the North Fork of the Flathead River; and west of Two Medicine and Cut Bank campgrounds on the park's eastern side.

The MOUNTAIN GOAT

For more than eight decades, the mountain goat has been the official symbol of Glacier National Park. It literally reigns at the very top of the Crown of the Continent ecosystem, and no animal better reflects the park's dizzying heights and rugged terrain.

Mountain goats (*Oreamnos americanus*) are monarchs of the high ramparts. They scramble with ease across alpine ledges that are inaccessible to most other mammals, including human mountaineers; they endure climatic extremes and fierce winds for most of their lives; and they give birth to kids on zeniths that jut into the clouds.

To classify the mountain goat as a goat, however, is a misnomer. It's actually a member of the Old World antelope family, and is in fact North America's only native "goat-antelope." According to biologists, Glacier's 1,400 to 2,000 goats are relatives of an elegant animal called the chamois, which is a famous fixture on many mountains in

Europe, including the Alps.

While mountain goats share the park's summits with bighorn sheep, the two species rarely co-mingle. Several physical characteristics distinguish them from one another. First, there is the goat's profile. Short and stocky, most goats have almost hunched backs and thick, shaggy, whitish-yellow fur. In contrast to the bighorn's curled horns, both male and female mountain goats wear dark horns that are dagger-like in appearance. The largest horns on a billy extend eight inches or more, while a nanny's spikes may be six inches or longer. Another distinctive feature of the mountain goat is its goat-like beard, which hangs from the throat like a mane. It's usually yellowish and oily.

The size of goats depends on the richness of their habitat, and whether they are hunted. Typically, males stand about 3 1/2 feet high and weigh between 120 and 250 pounds, while females are slightly smaller. Hunting is prohibited in Glacier, so some billies grow to be fairly large. As a result, rangers must constantly be on the lookout for poachers.

Perhaps the greatest reward for goat watchers is the opportunity to witness death-defying leaps. Quickly, one understands that the nimbleness of a mountain goat is no accident. The bottoms of their hooves contain a soft, almost leathery surface that provides traction on even the smoothest of rocks.

Male and females live separately until the autumn rut, or breeding season, when males vie for mates. Unlike bighorn sheep, however, mountain goats do not engage in ferocious head-butting. Some six months after mating, pregnant nannies give birth to as many as three kids, though the usual number is one.

Mountain goats inhabit high country that's beyond the reach of predators—with the exception of golden eagles. These birds have been known to drive kids off of cliffs to their death. The goats feed on staple foods of lichen, sedges, and alpine plants.

Unlike bighorns, they do not migrate downslope in a substantial way. However, wild goats occasionally leave their mountain fortress in search of salt, descending to lower elevations where mountain lions—their primary enemy—may be waiting. During one wildlife watching episode at a site called Goat Lick, a photographer recorded a mountain lion making a kill. Fortunately, nature has equipped goats with sharp hooves that often prove effective in repelling predators.

Although mountain goats are wary of people, they sometimes display high tolerance for human presence, and they can be seen crossing roads at various locations in the park or gathering at natural salt licks.

The MOUNTAIN GOAT

Where to look for mountain goats

Never approach or try to feed mountain goats. Be mindful of the animal's desire to be left alone.

➢ There are several excellent sites where you can view mountain goats from the roadside. Being crepuscular, Glacier's goats are most active in the hours around dawn and dusk. Logan Pass, at the crest of Going-to-the-Sun Road, offers several turnoffs that are top-notch places to spot goats as they scramble across alpine crags. Begin your day at the Logan Pass Visitor Center and ask a ranger for directions, or take a short hike down the Hidden Lake Overlook Nature Trail to look for goats on 8,760-foot Mount Clements and astride 9,195-foot Mount Reynolds.

➢ Goat Lick, located along Highway 2 south of Walton, is a gathering point for many goats in early summer.

➢ At Many Glacier, in the parking lot of the Swiftcurrent Motor Inn, photographers and park rangers routinely gather to look for bears, mountain goats, and bighorn sheep. Once you locate bighorn sheep, look higher up the slopes for mountain goats.

➢ In Waterton, goats are occasionally seen on the slopes above Rowe Lake. Seeing them will require a hike, though. To get to the trailhead, drive from the Waterton Townsite west along Akamina Parkway to the Rowe Lake trail.

 **The BIGHORN SHEEP**

Few acts of animal behavior are as dramatic as the head-butting displays performed by male bighorn sheep. For many hours every autumn, rams rear their bodies and, in a fitful display of passion, crack their racks against one another in order to win the right to mate. The rams' battle for dominance during the rut usually pits animals with similar-sized horns against one another. Mismatches are rare. This dangerous (and painful) ritual occupies just a fraction of a bighorn's life, but it's often the sight most vividly remembered by

The BIGHORN SHEEP

wildlife watchers visiting Glacier Park.

Bighorn sheep (*Ovis canadensis*) are expert climbers that materialize like sentinels on some of the highest crags in the park. Together with mountain goats, they dwell in a land of tundra that stretches to the extreme limits of the terrestrial environment—an unearthly world largely free of predators, but a place where the going is tough.

Identifying bighorn rams is simple. The rams sport a classic C-shaped curl resembling a corkscrew in their beige horns. The horns protrude from the skull just above the ears, and spiral toward a point. Unlike the antlers shed annually by members of the deer family, the sheep's bone-like growths are permanent. When watching females (ewes), notice that her spiked horns are much lighter in color than a mountain goat's.

Both sexes of bighorns tend to be larger than their mountain goat counterparts, weighing between 125 and about 300 pounds. The sheep are brownish-tan with a white rump patch, whereas goats are yellowish-white and much shaggier.

Bighorns have a stout, muscular frame that allows them to sprint across steep slopes with a low center of gravity. A sheep's hooves are critical to its survival, for they are hard-edged yet have a soft sole that provides traction. The sharp hooves can be used as an effective weapon against mountain lions that try to ambush bighorns or wolves that attack as the sheep migrate from the mountains to lower-elevation winter ranges.

The sociable female sheep congregate in small herds, which makes them easy to locate in the spring, when they form nursery groups to raise their lambs. The group is led not by a dominant ram, but by an elder matriarch. Rams roam separately except during seasonal migrations and the breeding season, when they join the larger herd. Healthy bighorns are relatively long-lived, reaching documented ages of 15 years.

Poaching of rams has been a chronic problem in many public wildlands throughout the West. Glacier adheres to strict federal regulations that outlaw hunting, but outside the park, where a few hunters are granted licenses to shoot bighorns in Montana, the opportunity is coveted. In 1987, Montana wildlife officials auctioned a single bighorn hunting license for $109,000, which indicates how eager some people are to hunt the animal.

Visitors can play an important role in protecting Glacier's resident population (between 300 and 500 bighorns) from illegal hunters and black market poachers. Report any suspicious activity to rangers immediately.

Where to look for bighorn sheep

Never approach or try to feed bighorn sheep. Respect their need for space.

➤ Bighorns are diurnal, meaning they're most active during the day. Many Glacier Valley, which sprawls beneath the Continental Divide, is a good place to start looking for them.

➤ Ewes and lambs can be seen on the slopes visible from the parking lot at the Swiftcurrent Motor Inn. Sheep have also been seen by hikers on the trail over Swiftcurrent Pass to Granite Park Chalet.

➤ In Waterton, you'll find a less rugged alternative at the Townsite during fall, winter, and early spring, or at the end of Red Rock Canyon Road northwest of Crandell Mountain. Waterton visitors are more likely to spot bighorns than mountain goats, and these are two of the best roadside spotting locations.

 **The MOOSE**

Speckled with 200 lakes and a backcountry nourished by a 1,450-mile circuit of streams, Glacier is a park in which water dictates the productive areas for wildlife watching. Where there's an abundance of ground moisture, the largest member of the deer family—the moose—cannot be far away. Moose (*Alces alces shirasi*) command attention because of their massive, gangling frames and the marvelous ornamentation of their antlers. The palmate spread of the antlers is what sets this species apart, but only bulls grow and shed their racks every year.

It's possible to mistake a cow elk for a cow moose, but closer examination reveals clear-cut differences. The cow moose's bodies are dark brown—almost black—when wet. Both males and females have a shoulder hump, an elongated snout, a bulbous nose, and a pendulous dewlap or fur "bell" that hangs beneath the chin. While both genders are massive, bulls occasionally reach a towering seven feet at the shoulders and weigh 1,000 pounds or more—a weight comparable to a horse.

Moose have extraordinarily long legs, allowing them to wade comfortably through river bottoms and marshes, constantly nibbling on leafy stems along the way. Willows are their staple food, and these trees are common in ponds created when beaver dams back up a creek.

The MOOSE

About 90 percent of the moose's winter diet is comprised of "browse"—the supple ends of twigs from conifers, willows, and shrubs.

Don't be fooled by the moose's docile appearance. Solitary bulls graze by themselves for a purpose, and cows with calves are notorious for chasing and occasionally injuring humans who venture too close. Be especially wary of bulls during the autumn rut, though this mating ritual is fascinating to watch from a safe distance. As a display of their masculinity, bulls stomp through meadows and flay grasses on their antlers. Occasionally, adult bulls tangle as they vie for territorial dominance. They drop their mighty antlers in late winter, but new ones begin growing almost immediately. The sheen of an antler is created in summer, when bulls scrape a soft layer of velvet off the hardening rack.

A healthy moose can live for 20 years, and in the protective sanctuary offered by Waterton-Glacier, it's not uncommon to see the same moose at the same wetland year after year. About 150 moose live in the park.

Where to look for moose

Never, under any circumstances, approach or feed moose. These animals are large and powerful, and can be extremely dangerous. Respect their desire to be left alone.

➢ Upper McDonald Creek, at the north end of Lake McDonald, is a fairly reliable site for seeing bull moose as well as cows and calves.

➢ The Many Glacier area, along the bottoms of Swiftcurrent Creek, is a prime spot for moose that are feeding on grasses and willows.

➢ Moose are seen along the North Fork Road, which parallels the North Fork of the Flathead River.

➢ In Waterton, moose are semi-regular visitors to the shores of Upper Waterton Lake near the Townsite. In winter, moose are sometimes seen along the Akamina Parkway.

The ELK (WAPITI)

Charles Russell painted elk in many wild retreats, but the forested slopes that rise above Glacier's Lake McDonald provided his favorite vistas. The legendary Western artist came often to the national park, and he was drawn on each visit to majestic herds of "wapiti." The term is a Shawnee Indian word for elk that aptly translates as "white rump" or "white deer."

The elk (*Cervus elaphus*) is the most common member of the deer family found in the park, and the males are distinguished by their elaborate antlers. Once a native species found from the Pacific Northwest to the Appalachians, the elk was completely extirpated from the East by overhunting. Today it primarily resides in the Rockies and the Pacific Northwest.

Autumn triggers frenetic changes for elk. It's the season in which bulls (males) are swept up in the rut. They initiate pitched battles for dominance, and try to recruit cows (females) into their mating harems. A group of cows led by a breeding male can number a dozen or more. The advent of the rut is normally preceded by the high-pitched "bugles" of bulls demarcating their territory, a memorable sound that may begin in late summer and last through early November. Feverish jousting sometimes follows as bulls go head to head, using their sharpened antler tines as offensive (and defensive) tools.

With the number of human visitors in the park diminishing and aspen leaves assuming festive colors, autumn is an ideal time to experience the drama of Glacier and its elk before the park shuts down for winter.

As the rut ends, elk move from the high country to lower-elevation ranges that provide enough accessible vegetation to sustain them through the winter. By March, bulls start dropping their antlers, and new branches start growing almost immediately. It should be noted that cow elk do not sprout antlers.

In late May and June, pregnant cows give birth to calves. At about this time, bulls are growing "velvet antlers," a term describing the soft surface of emerging antlers nourished by blood vessels and capillaries. The branches eventually harden, and bulls rub off the velvet overcoating in time for the rut.

Antlers can reveal information about a bull's age. As a general rule of thumb, bulls between two and five years old sport racks that contain only a couple of tines on either side. From the fifth year through the age of nine, bulls often carry six tines on each side. In the West, such an elk is considered a six-point bull, but observers east of the Mississippi River count the total number of tines and declare it a 12-point bull. Wildlife photographers in Glacier have been known to use artificial elk calls as a means of summoning bulls, but the practice is outlawed.

Rough estimates place the number of elk in Glacier at 2,200. You can easily identify an elk by looking for the following physical characteristics:

➤ A frame larger than a deer's but smaller than a horse's body. Bulls weigh between 500 and 1,000 pounds, and stand five feet tall at the shoulders. Cows weigh between 400 and 600 pounds.

➤ An elk's head is dark brown, the bulk of its body is tan, and the rump area is creamy white.

➤ The elk's tracks resemble cloven half-moons that are roughly four inches in diameter and rounder than a moose's tracks.

The elk's enemies prey particularly on elk calves, and include grizzly bears, mountain lions, and wolves. Elk are most active at night and during the twilight hours of dawn and dusk. Travelers getting an early start in the morning should scan the high-elevation meadows just below the tree line for groups of elk grazing on succulent plants in the summer months.

Whenever you see an elk near the roadway, proceed with caution, and never walk up to an animal that's grazing near your car. Cows are extremely protective of their calves, and bulls have been known to charge humans.

The ELK (WAPITI)

Where to look for elk

Even though elk may appear docile and tame, you should never approach or attempt to feed them. These animals are large, wild, and powerful, and they can be extremely dangerous. In addition, feeding wildlife is illegal.

➤ Look for elk at Lower McDonald Creek, between Apgar Visitor Center and West Glacier. Local herds pass through on their way to and from the Apgar Mountains and the higher slopes of the Flathead Range. They're visible in the autumn, when bulls can be seen (and heard) bugling as they joust in the heat of the rut.

➤ Look for elk in the high meadows across Going-to-the-Sun Road during the early morning hours of summer.

➤ At Two Dog Flat, located just below East Flattop Mountain and between St. Mary townsite and Rising Sun, elk gather on the fringe of a tree-lined meadow during the twilight hours of spring and autumn.

➤ One of the best places (in either park) to watch bulls bugling and jousting is in Waterton, near the intersection of Chief Mountain International Highway (Highway 17) and Highway 5 during September and October.

 **The MULE DEER**

The precipitous mountain environment of Glacier can impose demands on the large animals that migrate each spring from low-elevation valleys to high, broken grasslands. But mule deer (*Odocoileus hemionus*) are perfectly suited to the task. Brawny and durable, these roving creatures thrive on the park's rocky escarpment. "Mulies" are strictly a Western phenomenon, and are larger evolutionary offshoots of the widely distributed white-tailed deer.

It's easy to distinguish between bucks (males) of each species. Darker brown hair covers the backs and foreheads of mule deer, though the mule buck's truly distinguishing features are a conspicuous patch of white inside its large ears and a black-tipped tail that protrudes from a cream-colored rump. The territories of the two species rarely overlap, except in winter. Whitetails tend to be creatures of moist river bottoms, while mule deer exist chiefly in drier evergreen forests. However, both mule deer and white-tailed deer can

The MULE DEER

be found throughout Glacier.

When fleeing, mule deer lope along in bounding strides. All four feet touch the ground at once, and their dark tails hang down. Whitetails, on the other hand, sprint with their white tails flagged and flared. The mule deer's means of evading predators would be costly over flat, treeless country, but its running style, called "stotting," is effective in mountainous terrain. Note in the illustrations provided how the tracks of mulies and whitetails differ.

The antlers on a mule deer buck are significantly larger than those on a male whitetail, measuring up to four feet across. These spreads are dropped in mid-winter, and new sets begin growing soon thereafter.

Mule deer subsist on a year-round diet of brushy vegetation. Nomadic for much of the year, the deer will sometimes assemble on winter ranges in a process known as "yarding up."

In May, does give birth to fawns (usually twins) after a six-month pregnancy. Glands situated near the hooves on the female's hind legs emit an odor that allows fawns to recognize their mother. The primary predators of mule deer in the park are mountain lions, wolves, and coyotes.

Where to look for mule deer

Although mule deer may appear tame, never try to approach or feed them. They can be aggressive and dangerous. Feeding wildlife is illegal, and it leads to habituation that harms the animal in the long run. If you see others feeding deer, encourage them to stop; you'll be doing the animal a favor.

➤ Look for mule deer at Many Glacier Road, between Swiftcurrent Motor Inn and the town of Babb 12 miles away. Mule deer are often visible in this area during the early morning hours and in the evening just before sunset.

➤ The Hidden Lake Nature Trail, accessible from Going-to-the-Sun Road, is located within the mule deer's summer range. In July and August, you can view bucks from the boardwalk as they forage on juniper, berries, and other plants native to the tundra environment.

➤ You'll find a resident population of mule deer along Lower McDonald Creek, between Apgar Visitor Center and West Glacier. These deer inhabit the foothills of both the Apgar Mountains and the larger Flathead Range.

➤ In Waterton, look for mule deer along the Red Rock Canyon Road, in the hills above Blakiston Creek.

The WHITE-TAILED DEER

As these words appear in print, white-tailed deer are themselves creating a chapter in the natural history of Glacier Park. Although whitetails (*Odocoileus virginianus*) probably inhabited the park thousands of years ago, they were replaced in more recent times by the hardier mule deer. Yet, on the far western fringes of the park, whitetails are rapidly re-colonizing moist river bottoms, offering wildlife watchers a plethora of new opportunities and providing prey for wolves and other predators.

According to biologists, the white-tailed deer is the oldest of all the deer species in the western hemisphere. It has managed to cope so

The WHITE-TAILED DEER

well with habitat fragmentation that whitetails now thrive in environmental settings as diverse as suburban lawns and wilderness areas. In Glacier, the Continental Divide influences deer habitat. West of the Divide, where moisture is trapped and prevented from reaching eastern sections of the park, whitetails have taken to river-based corridors lined with willows, aspen, and cottonwoods. However, both white-tailed deer and mule deer are found throughout Glacier.

Whitetails are leaner and more reddish-brown than mule deer. Of course, the surest clue in identifying a whitetail is the white underside of the tail, which flashes when the animal flees or becomes excited. Whitetail bucks sport smaller antlers than mule deer. The antlers are shed in winter, then replaced in summer. Whitetail fawns are born in late spring, following a seven-month pregnancy. The first time she gives birth, a doe usually produces a single baby, but she'll yield twins on subsequent deliveries.

Whitetails are not particularly social creatures. They may congregate in groups on winter ranges, but they have no affinity for herding. At many locations, though, they tolerate humans as they ruminate and dip to munch on food. Note that whitetail tracks are rounder than mule deer prints.

The primary predators of whitetails are wolves, mountain lions, and grizzly bears—generally in that order.

Where to look for white-tailed deer

Never approach or try to feed whitetails. The animals are not tame, and they're potentially dangerous.

➤ Nearly the entire western half of Glacier National Park provides excellent habitat for whitetails. Because they're crepuscular and nocturnal, whitetails are routinely seen during the early and late daylight hours. Drive the Camas Road between Apgar and the park entrance at Camas Creek.

➤ Another promising option is the gravel North Fork Road (closed in winter) that adheres to Glacier's western boundary along the North Fork of the Flathead River and all the way northward to Kintla Lake, just south of the Canadian border. Along this route, consider taking the dirt road into Bowman Lake, where whitetail sightings are frequently reported.

➤ In Waterton, whitetails are regularly sighted near the shores of Cameron Lake at the end of the Akamina Parkway, and along the Chief Mountain International Highway (Highway 17) along the Belly River.

 **The BISON (BUFFALO)**

Though they're emblems of the frontier, free-roaming bison (*Bison bison*) have been absent from Glacier for more than a century. Historically, small groups of bison traveled seasonally up the lower-

The BISON (BUFFALO)

elevation drainages of both parks. Better known by their colloquial yet incorrect name, "buffalo," bison are classified as bovids, or members of the cattle family. In fact, they're the only species of cattle native to this continent.

Male bison (bulls) and females (cows) both have permanent horns that are not shed. The animal's coat consists of shaggy, reddish-brown fur, but the head itself is dark brown. Some individuals have a dark brown mane and beard as well. Bulls grow to a height of six feet at the shoulder, while females reach five feet.

A small herd of the wild bovines is enclosed in Waterton Park, at the famous Bison Paddock. These animals are part of an estimated 30,000 bison that survive in parks and wildlife refuges across North America. The enclosure gives visitors an opportunity to experience the 2,000-pound behemoths from the safe confines of a car. Remember that visitors cannot leave their vehicles, because even semi-domesticated animals can become riled by humans who approach them too closely.

Where to look for bison

➤ Visit Bison Paddock, a drive-through bison enclosure just off of Highway 6 at the northeastern edge of Waterton Park.

 **The WOLVERINE**

The wolverine (*Gulo gulo*) has earned its reputation as a brawler and a scrappy stalker of wildlife carcasses. A product of the boreal forest and Arctic tundra, the wolverine is rare in the lower 48 states because such habitat is uncommon.

The chances of encountering one of these reclusive animals in Glacier are slim. When a wolverine is spotted, this largest member of the weasel family is often mistaken for a small bear as it moves across snowfields or through the dense understories of fallen trees. Wolverines have been known to drive mountain lions, coyotes, and even grizzly bears away from a fresh carcass—behavior that illustrates the animal's ferocity.

Despite its relatively small size, the wolverine attracts a great deal of attention. Its hulking, bear-like frame rides low to the ground. The largest males weigh about 60 pounds. To identify wolverines, first note their dark brown coat, which is bordered with a band of yellowish fur across the forehead and back, extending to the rump area. The wolverine's other distinguishing features include long claws that emerge from all four paws and a short, bushy tail.

Males interact with females only long enough to mate. They then embark on long solo journeys across their territory, which can encompass several hundred square miles. Wolverines can swim, climb trees, and negotiate deep snow. Rather than burying a carcass for later consumption, the wolverine will spray the carrion with its musk in order to keep other predators away.

On the rare occasions when humans have come across wolverines, the animals were often heard growling to themselves as they walked. Note that wolverines should never be approached. If, on the other hand, a wolverine approaches you, it's best to back away and give the animal a wide berth.

Where to look for wolverines

➤ Appearances by wolverines are difficult to predict. However, the animals are sometimes sighted at the edge of forests along the Many Glacier Road; in the snowfields visible from Going-to-the-Sun Road over Logan Pass; along the shores of Lake McDonald; and along the northernmost reaches of the North Fork Road within Glacier (south of Kintla Lake).

 **The BADGER**

Badgers (*Taxidea taxus*) are best viewed from a long distance away. Fiercely territorial, they will charge virtually any intruder that approaches their dens or food sources. Because badgers in Glacier are protected from harassment and hunting, the animals abandon their usual nocturnal foraging habits and are often visible during daylight hours.

The markings on a badger's coat are striking, and the design of its body is imposing—even though it's a creature that weighs less than 30 pounds. You'll see a conspicuous line of white under the the badger's snout, across its flattened head, and along its upper backbone. It also has a patch of white fur on both cheeks and ears, broken only by a hook-shaped pattern of dark brown. Most of a badger's back is marbled gray and brown.

A carnivore with extraordinary strength for its size, this scrappy member of the weasel family spends most of its life digging—making a new den, escaping from one of its few enemies, or bulldozing its way into a ground squirrel den for a meal. The badger has a low, tank-like frame with muscular forearms and curved claws. It prefers residing in treeless meadows not too far from the dens of its prey.

Coyotes are one of the few rival predators that badgers tolerate, and that's only because the two have developed a partnership. As a badger digs its way into the den of a squirrel or chipmunk, a coyote will wait outside the den's escape hatch with open mouth. Despite the badger's habit of simply leaving the bones and hair of its victims outside the den, it's fairly fastidious about grooming its coat, and it buries its own scat.

Where to look for badgers

Although small, badgers are aggressive and dangerous. Never approach or try to feed them.

➤ There are established badger den sites along the road between the west park entrance and Apgar Visitor Center just north of the road that crosses Lower McDonald Creek.

➤ Look for badgers along the eastern boundary of the park south of St. Mary on Highway 89.

➤ Badgers are regularly spotted in the vicinity of Two Dog Flats on the eastern border of Glacier, and in the Big Prairie area.

➤ In Waterton, badgers are sometimes seen inside the Bison Paddock. A badger den has also been located in the Waterton Townsite campground in recent years. In addition, there are badgers near the junction of the Chief Mountain International Highway (Highway 17) and Highway 5.

The BEAVER

A builder of empires, the beaver (*Castor canadensis*) is an aggressive landscape architect whose backwater creations fill an important niche in nature. By damming streams and thereby creating ponds, the beaver spawns habitat that benefits some of Glacier's most popular animals.

Willow trees thrive on the edge of a beaver's reservoir, providing an important source of food for moose, elk, and deer. Even ruffed grouse gather in the deciduous trees that spring up around the beaver's pond, and waterfowl find it a welcome stopover.

With its sharp bucked teeth, the beaver can chew the trunks of hardwood trees to a conical point, then fell the entire tree. It uses its flat, paddle-like tail to navigate in the water. The beaver's brown, shiny fur insulates it from the icy water that confronts it when it emerges in winter from hatches beneath its spherical lodge.

Most biologists consider the beaver a constructive force in altering the environment, but there are ranchers outside the park who resent the animal's ability to flood pastureland or stop the flow of a creek, so that a formerly irrigated field of cropland perishes. In places where beavers are viewed as a nemesis to agriculture, they have been shot or removed from the area.

Across most of the West, beaver populations have been rebounding for almost a century after extensive trapping nearly wiped them out. The inland waters of Glacier have been an important sanctuary, because animals that unknowingly swim beyond the park border may be trapped and killed for their pelts.

If a beaver is startled by human presence, the animal will slap its tail down hard on the water's surface and then dive out of sight.

Where to look for beavers

Never approach or harass beavers. Give yourself and other visitors the pleasure of observing them as they naturally behave.

➤ Lower McDonald Creek between Apgar and West Glacier has an old beaver lodge, uninhabited but still intact. Upstream in Lake McDonald, swimming beavers are seen regularly, especially in the early mornings and late afternoons.

➤ Both the North Fork of the Flathead River on the park's western boundary and St. Mary Lake on Glacier's eastern edge are good beaver-watching sites.

➤ In Waterton, beavers are occasionally seen at picnic areas between Middle and Lower Waterton lakes.

 ## The PORCUPINE

Porcupines (*Erethizon dorsatum*) are known as the prickly pears of the animal kingdom. These shy mammals are rare in Glacier, though their population varies greatly over time.

Although it holds up to 30,000 spiny quills in its quiver, an adult porcupine—contrary to popular belief—cannot throw or "shoot" the quills. Rather, these quills are earned by whomever puts an unwanted nose into a porcupine's defensive armor. Although the quills can easily be removed from a porcupine's coat, it's much more difficult to remove them from a victim. The barbs at the end of each quill result in painful impaling. Predators of the porcupine within Glacier include coyotes, mountain lions, bobcats, and fishers, though none of these count the porcupine as a primary food source.

Shaggy and yellowish-brown, porcupines have soft, non-menacing faces and bulky frames that wobble as they walk. They're adept at

The PORCUPINE

climbing trees. Males are solitary until the autumn mating period begins, when they join females and engage in high-pitched squealing during courtship. Young porcupines are born with a full set of quills that are soft at birth but harden within hours.

One sign of porcupine presence is their trademark paw prints, made by soles that look as if they're covered by cobbled pebbles. Usually, marks from their long claws are also visible.

Another sign is a tree that has its bark pulled away from the trunk and tooth marks left in its place. The animal eats bark, and complements its vegetarian diet with twigs, leaves, lupines, and other plants.

Unfortunately, the porcupine also has a taste for salt that's sprinkled on roads to melt ice. Glacier does not salt its roads, but eating salt has become a mortal liability elsewhere for the porcupine. Each year, porcupines are struck and killed by vehicles on highways. The animal's natural lifespan is seven to eight years.

Where to look for porcupines

Never approach porcupines or try to feed them. If you're camping with your dog, make sure it's restrained. A dog with a muzzleful of porcupine quills is a mournful sight.

➤ Because they're nocturnal animals, porcupines are occasionally seen waddling along park roads just after sunrise and in late evening. Drive with caution.

 ## The RIVER OTTER

There are no creatures in Glacier more delightful to watch than river otters (*Lutra canadensis*). They're the second-largest marine mustelid in North America (only sea otters are bigger), and their long, narrow frames are typical of members of the weasel family.

Excellent swimmers, river otters reach lengths of three to four feet. With their skinny, furry tails serving as rudders, otters use their webbed feet to propel themselves and a valve system to keep water from going up their noses.

Park visitors warm easily to the otter's playfulness, whether it's sliding down snowbanks headfirst into rivers, or swishing alongside canoeists who venture across Glacier's larger lakes. While comparisons to humans seem anthropomorphic, scientists say otters actually appear to maintain a childlike joy in their daily routines and family outings.

The RIVER OTTER

Warmed by a coat of brown fur, otters are carnivorous waterfarers highly adept at catching trout. They're quite gregarious, and they sometimes navigate in pairs. Males weigh about 25 pounds, females slightly less.

Otter presence is indicated by their five-toed tracks in the mud, and by greenish scat laced with fishbones or scales. The animals dig permanent dens into the banks of streams or shoreside mounds, accessible through underwater lanes called "runs." Dug deep into the lake or river bottom, these passageways don't usually freeze solid during winter, so otters have a safe, year-round route between warm shelter and open water.

Otters inhabit Glacier today, but at one time fur trappers had nearly extirpated them from watersheds in the Rockies. Otters are protected in the park, so there's a stable population that's active both day and night.

Where to look for river otters

Never approach or feed otters. Though they seem friendly, they have been known to bite people who venture too close.

➤ Otters are sometimes visible from land or during boat excursions on Lake McDonald and Lower McDonald Creek.

➤ Look for otters along the Middle Fork of the Flathead River, which is paralleled by U.S. Highway 2.

➤ Otters can be seen along most of the North Fork of the Flathead River, running along the west park border from Polebridge southward to the North Fork's confluence with the Middle Fork.

BIRDS

 The BALD EAGLE

The BALD EAGLE

For many of us, bald eagles are symbols of a clean environment and in both parks these majestic flyers are relatively common. Glacier and its neighboring environs are found along a major north-to-south bald eagle migration corridor. Eagles heading to more southern areas of the U.S. prior to the onset of winter often make pitstops in park waters. The sight of these national symbols is inspiring in any season. Over the years, several nesting pairs of bald eagles have also made their home in and around the park.

Bald eagles represent more than liberty and rugged individualism; these birds of prey are indicators of a clean and healthy environment. More than meat-eating predators, bald eagles are useful and beautiful scavengers that take advantage of dead fish and carrion. Large lakes like Lake McDonald, girded by the tall conifers in which eagles build nests, provide a measure of the isolation the birds prefer.

Eagles attract attention by virtue of their distinctive profile rather than their calls. Bald eagles are not truly bald. Their name is derived from the Greek word *leucocephalus*, meaning "white-headed." The bird's crown and tail feathers take on the famed snow-white tint only when a bird reaches adulthood, and the process of maturation may take four or five years and involve five different molts (feather sheddings). In the meantime, observers unfamiliar with the markings of immature bald eagles may confuse them with golden eagles, which are also common in Glacier.

At adulthood, a bald eagle's wingspan may reach between six and seven feet. When it's not flying, the bald eagle seeks a high perch in a conifer or cottonwood tree, from which it can scout its surroundings. In this sitting posture, the bird is 2 1/2 to 3 feet tall.

Native only to this continent, bald eagles were spiraling toward extinction in the lower 48 states following decades of spraying farmlands and forests with the pesticide DDT. By 1963, fewer than 500 pairs of bald eagles were observed in all of the contiguous U.S. Today, though their numbers are increasing, balds are listed as an endangered species in 43 states.

Whether walking on the shore of a lake or canoeing its waters, humans should take special care to prevent disturbance of bald eagle nest sites. During the late winter and spring, most sites are placed off-limits to human intrusion by park officials. Use a spotting scope or

high-powered binoculars, and observe bald eagles and their nests only from a distance. Intrusion may cause adult birds to abandon their young, or may interrupt adults that are trying to feed the eaglets. When you're making plans to scan the skies and treetops, remember that bald eagles are most active during daylight hours.

Where to look for bald eagles

Never approach bald-eagle nests.

➢ Look for bald eagles at Lower McDonald Creek, between Apgar Visitor Center and West Glacier. Historically, this was once a major point of convergence for the hundreds of bald eagles that came here in the autumn to fish for spawning salmon. A few resident birds remain, and they can be seen soaring over the southern end of Lake McDonald or perched in treetops along lower McDonald Creek. Every autumn, dozens of migrating eagles stop here on their way southward from Canada, but it's difficult to predict their arrival.

➢ Bald eagles are sometimes seen flying over St. Mary and Lower St. Mary lakes on the eastern boundary of the park and Bowman Lake in Glacier's northwest corner.

➢ In Waterton, look for bald eagles in the early morning hours over Upper and Middle Waterton lakes outside the Townsite.

 ## The OSPREY

The osprey (*Pandion haliaetus*) is often called the "fish eagle," because it dives after trout and salmon even more aggressively than the bald eagle. Sustained solely by a diet of fish, these carnivorous birds seek solitude along the far shorelines of larger lakes.

The most overt indicators of osprey presence are their nests, piled with twigs and branches at the tops of evergreens or cottonwood snags. Each spring, breeding females return to the same nests, adding more sticks after a long migration from winter retreats in Central America. The females usually lay a clutch of three eggs, and human disturbance—particularly during incubation—should be avoided at all costs.

Although osprey and bald eagles perch and soar over waterways, the two birds of prey are quite different in appearance. The heads of both the male and female osprey are white with a dark brown band running across the eyes. By comparison, the bald eagle's head is entirely snow-white. The top of an osprey's body and tail is generally dark-colored, while the bottom is mostly white. An adult osprey's wingspan reaches about 60 inches across, compared to the bald eagle's wingspan of 72 inches or more.

It's a treat to watch an osprey float quietly above a lake or river, then suddenly dive legs-first toward the water. Following impact, an osprey will often pull itself into the air with a fish impaled on its claws. To aid in their angling, osprey have talons equipped with toes and coarse footpads that allow them to maintain a firm grip on their slippery catch.

Where to look for ospreys

Never approach an osprey nest.

➤ Look for osprey on Glacier's bigger lakes, such as Lake McDonald, St. Mary Lake, and Sherburne Lake.
➤ In Waterton, both Upper Waterton and Middle Waterton lakes contain summer nesting populations of osprey. There is also an osprey nest at the park entrance, just inside the boundary.

The GOLDEN EAGLE

Dwellers of isolated valleys, open savannas, and mountain canyons, golden eagles are much more predatory than their bald eagle relatives, usually killing their own prey rather than scavenging. This species of giant raptor is named for its "golden" (actually dark brown) plumage, but what really distinguishes it from the bald eagle is that it hunts primarily over land, rather than water.

Golden eagles (*Aquila chrysaetos*) are an imposing presence in the sky as they soar high above meadows, using their keen eyesight to spot rabbits, rodents, and even deer fawns on the ground. They're slightly smaller than bald eagles, though their wingspans can reach more than six feet. Accelerating into a downward dive that can exceed 160 miles per hour, the golden's wing strength allows it to maneuver gracefully while nabbing its prey. Occasionally, however,

golden eagles perish as a result of crashing into the ground or power lines (outside the park). Power line electrocution also occurs when eagles use the poles as perches.

The golden eagle population declined sharply in the West during the 1960s, following decades of exposure to the pesticide DDT—which was then banned in the early 1970s. Since the ban was imposed, a noticeable recovery has taken place. Goldens make their nests in cliff walls, and some sites have been occupied by several generations of birds.

When identifying golden eagles, look first to the color of the feathers, then to the head. Although they're mistaken for immature bald eagles, remember to consider the type of terrain, for bald eagles rarely leave lakes or streams. From the ground, the outline of a golden resembles that of a hawk. A patch of white is visible on the fanned, grayish tailfeathers and between the brown and gray of the wings. It may take four years and two or three moltings (feather sheddings) before the adult bird's full plumage appears.

Despite their reputation among sheep ranchers for raiding young lambs, goldens will hunt wildlife if the prey is available. This eagle species, while rare, is not listed as either threatened or endangered under the Endangered Species Act, but is instead protected under the federal Migratory Bird Treaty and the Eagle Act.

Where to look for golden eagles

➤ Look for soaring golden eagles along the upper stretches of Going-to-the-Sun Road.

➤ In Waterton, goldens are occasionally seen near the cliffs of Red Rock Canyon.

The RED-TAILED HAWK

Exceptional eyesight grants the red-tailed hawk a commanding view of Glacier's grasslands. On the eastern fringe of the national park, there's an unmistakable influence from the high plains as sagebrush meets the mountainous, coniferous foothills. The red-tail (*Buteo jamaicensis*) is one of the most identifiable raptors in the contiguous U.S., and it's the most abundant hawk in the park.

In Glacier, visitors see the western variation of the red-tail—a moderately large, solid-brown soarer that has the characteristic rufous

The RED-TAILED HAWK

(reddish) tail. From the ground, the hawk appears light-colored with spots of brown on its breast. Its wingspan reaches four feet or more.

Unlike golden eagles and peregrine falcons, red-tails are tolerant of humans and frequently conduct their aerial hunting along roadsides. Typically, a red-tail will land in a tall pine and scout out its prospects. It may soar into the sky, then plunge into a glide parallel to the ground that allows it to spot unsuspecting mice, voles, rabbits, and ground squirrels. The red-tail is comfortable in open savanna and forested country at middle elevations. Like eagles, red-tails often mate for life.

Where to look for red-tailed hawks

➣ The eastern edge of the park near Two Medicine, the Cut Bank campground, and St. Mary provide exceptional habitat for red-tailed hawks.

➣ Red-tails are found also in the rugged river meadows of the North Fork of the Flathead, along the western periphery.

➣ In Waterton, red-tails are spotted near the Belly River, along the Akamina Parkway, and along the Red Rock Canyon Road.

 # The COMMON LOON

The calls of a loon are nothing short of the audio embodiment of wilderness. Although Glacier and Waterton lie near the southern edge of the common loon's summer range, the tranquil waters and remote lakes found here attract a regular contingent of these large, handsome waterfowl.

Common loons (*Gavia immer*) are large and elegantly marked. Their backs and wings are almost black, with whitish spots. Their necks are dark green, highlighted by a striped ribbon of white, and their heads are dark with a dark bill. The loons that inhabit Glacier migrate from wintering areas on the coast. They are among four different species in the loon family. Native to most of Canada's boreal evergreen belt, the common loon prefers to live in secluded lakes

The COMMON LOON

away from human civilization. However, people are attracted to its vocalizations. The loon's most distinctive calls are the tremolo, the yodel, and the wail, and each has a different meaning.

One characteristic that set loons apart is their penchant for making deep dives to find fish and aquatic plants. They disappear beneath the water's surface, startling observers when they emerge hundreds of feet away.

Where to look for common loons

➤ Loons are often present on the isolated waters of Bowman, Kintla and Quartz lakes, which are accessible from the North Fork Road. Campers hear their calls in the early morning and evening hours.

➤ In spring, just after the ice has left, and in October, you can spot loons on Lake McDonald.

➤ In Waterton, loons are found on Upper Waterton Lake, particularly after the tourist season has slowed down.

The HARLEQUIN DUCK

Normally, when we envision migrations, a south-to-north axis comes to mind. But for harlequin ducks (*Histrionicus histrionicus*), the strenuous flight is a west-to-east journey that carries these secretive birds to Glacier's streams from the sea. Each year, from their wintering areas along the Pacific coastline of Washington and British Columbia, harlequins migrate inland to breed and raise their young. However, the birds are rare.

Glacier Park and other hinterlands along the Rocky Mountain front represent the far fringes of the harlequin's summer range, though the area's clean, fast-moving brooks are of critical importance because they provide habitat and food during the breeding season.

Harlequins are just one of 43 different waterfowl species that breed on the North American continent, but none is more beautiful. Vividly colored, harlequin drakes (males) are rivaled in aesthetic appeal only by drake wood ducks. The males have a slate blue body streaked by white, and chestnut plumage across the head and wings. In their coloration, harlequins are often thought to resemble circus clowns. Hens (females) are brownish-gray, with white streaks on their heads. The females build their nests among tangled tree roots or boulders along streams with a high gradient. The hen generally lays a clutch of six cream-colored eggs.

Seeing a harlequin in Glacier or in the adjacent Bob Marshall Wilderness is a memorable but uncommon experience. The birds arrive as soon as the ice thaws and leave in early autumn. Harlequins eat aquatic invertebrates such as caddis and stone fly larvae that are present only in swift-moving streams. Their main predators are river otters and mink.

Where to look for harlequin ducks

Never approach harlequins or their nest sites. They are extremely sensitive to intrusion.

➤ Upper and Lower McDonald Creek, to the north and south of Lake McDonald, is known as a haven for nesting harlequins, but the birds are secretive and sensitive. Disturbance should be avoided at all costs.

➤ In Waterton, harlequins are occasionally seen along the portion of Red Rock Canyon Road that parallels Blakiston Creek, between Crandell Campground and the road's end.

The WHITE-TAILED PTARMIGAN

White-tailed ptarmigan (*Lagopus leucurus*) are models of adaptation, and are admired by ornithologists as feathered chameleons. Capable of modifying their plumage so it reflects seasonal changes in the environment, ptarmigan have found a home in Glacier's mountain passes.

Scientists say that ptarmigan (pronounced *tahr-mah-gun*) are relics from the Ice Age, thriving where alpine plant communities are present in a tundra environment similar to Alaska and northern Canada. In fact, the white-tailed ptarmigan is a close (but smaller) relative of the rock and willow ptarmigan that abound in the northern latitudes of the continent. Glacier's population of white-tailed ptarmigan is similar in size to other remnant colonies in the Sierra Nevada Mountains, the northern Cascades, and isolated hamlets of the Rockies.

Despite their soft, fragile appearance, ptarmigan are hardy birds that remain on the frigid plateaus throughout the winter. On the coldest days, they burrow into the snowpack, which insulates them from the bitter elements.

In summer, ptarmigan blend with their rocky habitat by assuming brown, mottled plumage set off by white bars on the tail, breast, and wings. But as the days shorten, their feathery frames begin turning the pearly white that will serve them well as winter camouflage.

One of ten different species of grouse native to North America, ptarmigan will fly longer distances when flushed than their woodland cousins, though they actually prefer to run when trying to evade intruders. Thick feathers on their feet enable them to trot across snow. Their ground-foraging habits, however, make them vulnerable to predation by red foxes, coyotes, grizzly bears, and birds of prey.

Look for white-tailed ptarmigan above the tree line in wildflower meadows and near rock outcroppings.

Where to look for white-tailed ptarmigans

Never approach or try to feed ptarmigan.

➤ In a productive year, a dozen or more ptarmigan are seen along the boardwalk at Hidden Lake, which is accessible from the Logan Pass Visitor Center on Going-to-the-Sun Road.

➤ Hikers bound for Granite Park Chalet near Swiftcurrent Pass, or Sperry Chalet on the Gunsight Pass Trail, have spotted ptarmigan at higher elevations above the tree line.

➤ In Waterton, look for ptarmigan above the Townsite on the hiking trail leading to Alderson Lake.

A WATERTON-GLACIER GALLERY

Snowshoe Hare
Lepus americanus

It may look like just another rabbit with oversized feet, but it plays an important role in Glacier National Park. The snowshoe is nearly the sole food staple of lynx. It's a major ingredient in the diet of wolves. And its own population swings affect the abundance of half a dozen other predators, including owls and hawks. Gifted with broad, snowshoe-like hind paws, this hare can dash across deep snows at speeds of up to 30 miles per hour. Although it's stalked by many different carnivores, the big-eared, three-pound hare uses tricks to elude predators. In summer its furry coat turns brown to provide camouflage and in autumn, just before the onslaught of snow and after a molting period, the emerging new fur is white. As a further means of cloaking its activities, the shy snowshoe is active only at night. The population of snowshoes in Glacier peaks and crashes on about a ten-year cycle. Look for the snowshoe hare in the vicinity of Many Glacier, the meadows around the Cut Bank campground, and in Waterton near Belly River.

Muskrat
Ondatra zibethicus

A hollowed mound of mud, cattails, and small sticks, the muskrat's home looks like a miniature beaver lodge. In a pond, this mound is the surest indication of muskrat presence, while in streams with a swift current, muskrats burrow into the bank from underwater passages. Dark brown with a long, somewhat flattened and leathery tail, muskrats use their webbed hind feet to propel them through the marine environment. They're active mostly at night, and they subsist in Glacier on a wholly vegetarian diet. Wildlife watchers usually find them in wetlands created by beavers.

A Waterton-Glacier Gallery

Mink
Mustela Vison

Where there are muskrats, wildlife watchers usually spot one of the muskrat's fiercest predators—the mink. Larger than a weasel but slightly smaller than the pine marten, the mink measures up to 2 1/2 feet, with a long tail and chocolate or reddish-brown fur that's prized by trappers. The mink is a capable swimmer, and its ability to flee on either land or water makes it less vulnerable to predators. Look for mink on the southern shores of Lake McDonald and along McDonald Creek.

Short and Long-tailed Weasel
Mustela frenata/ Mustela erminea

Neurotic, curious creatures, weasels are found along many waterways and forested edges in the park. Pound for pound, they're among nature's most formidable predators, weighing eight ounces or less but preying on animals twice their size. The weasel's coat is brown with light tan belly fur during the summer, but it changes to white in winter, reflecting the landscape. You'll find the long-tailed variety of weasel along most middle-elevation rivers and lakes. Short-tailed weasels, also called "ermine," are sometimes seen near campgrounds and along wooded trails.

Pine Marten
Martes americana

The pine marten resembles a brown-phase weasel, but it's a larger animal with a tail that covers a third of its two-foot body length. The marten is a tree-dwelling resident of old-growth forests, finding shelter under dark, coniferous canopies. It's elusive, but wildlife watchers occasionally spot one along the shore of Lake McDonald, near the Sun Point Nature Trail on St. Mary Lake, or on hiking trails off of the North Fork Road.

Fisher
Martes martes

The fisher is a creature that thrives in the deep boreal forests that dominate Canada. It has all the classic trademarks of a mustelid—a long, thin, strong frame, a bushy tail, and short legs. In fact, a fisher looks like an oversized mink. Adults can attain weights of 12 to 18 pounds and lengths of over three feet. Part tree-climber, part swimmer, part burrower, the fisher lives in undisturbed backcountry and is extremely rare in the contiguous U.S. You're not likely to spot a fisher from the highway.

Pika
Ochotona princeps

Scree slopes and rockslide areas are home to the famed pika of the high country. The animal migrated to this continent across the Bering Strait from Asia thousands of years ago. Although these tiny mammals are rabbit-shaped, they also resemble pet-shop guinea pigs. The pika has reddish-brown fur; wide, rounded ears; white paws; and no tail. It sounds a shrill, high-pitched bleat (or whistle) when it's threatened. In Glacier, the bleat serves to inform humans of pika presence, but it also tips off birds of prey and weasels—the animal's main predators. You'll find pikas at some pull-overs along Going-to-the-Sun Road ascending Logan Pass, and along the Red Rock Canyon Road in Waterton. Do not feed these animals.

Hoary Marmot
Marmota caligata

The hoary marmot goes by many names. Some folks know it as the "rockchuck" or the "mountain marmot," but visitors traveling the roadside of Glacier will remember it as a chunky rodent that packs a pretty good whistle. The hoary marmot is silver-gray and haunts rock piles in alpine terrain. Marmots can be fairly large (up to 20 pounds), and they may be persistent in begging for human food. Resist the temptation to give in, for they have been known to bite, and dependence on table scraps hurts them in the long run. The primary predators of marmots in Glacier are grizzly bears, coyotes, badgers, and birds of prey. From mid-October until February, these marmots disappear from sight and hibernate in underground nests. Look for them along the Hidden Lake Trail boardwalk leading from the Logan Pass Visitor Center, and near Goat Lick in the park's southern end.

Columbian Ground Squirrel
Spermophilus columbianus

Common along the Hidden Lake Nature Trail that leads out from the Logan Pass Visitor Center, Columbian ground squirrels make a brief summer appearance, then head back into their dens to hibernate for eight months. The attention-grabbing squirrel has a golden face and belly, a mottled, orange-gray back, and a bushy tail. Native only to a narrow belt of mountains from central Idaho to northern British Columbia, the Columbian ground squirrel stuffs itself with seeds, tubers, and plant stems as well as insects. It's found throughout Glacier. Never approach or feed ground squirrels. Feeding park wildlife is illegal, and these animals have been known to bite.

Red Squirrel
Tamiascurus hudsonicus

Like the Clark's nutcracker, the red squirrel has an important relationship with Glacier's whitebark pine trees, and greatly influences the foraging habits of federally protected grizzly bears. As soon as the seeds ripen in whitebark pine cones every autumn, red squirrels ascend into the trees and compete with the nutcrackers in harvesting seeds. Dropping entire cones to the ground, red squirrels (also known as pine squirrels) store the cones in underground chambers for later use. Grizzly bears find these nutritional treasure troves and gorge themselves on seeds, a practice that helps them add enough fat to survive winter dormancy. Red squirrels are found throughout the park. Never feed either red squirrels or chipmunks. Feeding park wildlife is illegal, and you may hurt the animal's chance of survival by turning it into a beggar.

Peregrine Falcon
Falco peregrinus

Less than 25 years ago, the peregrine falcon appeared to be nosediving toward extinction. Across North America, populations of these striking birds of prey were declining, due in large part to use of the pesticide DDT. Even today, the peregrine remains the rarest of 15 species of hawks that were once distributed widely across North America. The National Park Service has joined forces with the Peregrine Fund to manage and reintroduce this endangered species. In recent years, Glacier visitors have reported sightings in the rugged cliffs rising from St. Mary Lake, but researchers say there are no known peregrine nest sites in the park—current or historic. Male peregrines can be identified by their intense features—dark gray head feathers that resemble a helmet, pointed "falcon" wings and tail, and occasional dark streaks on the white chest. Females are a darker brown on top and bottom.

Canada Goose
Branta canadensis

Because the Canada goose is one of the best-known wild residents of Canada, you'd expect it to be abundantly represented in at least half of Waterton-Glacier International Peace Park. However, there's not much habitat here that's suitable for these large, black and gray birds. The lakes are too deep, many of the rivers are too swift, and there are no agricultural lands to provide geese with grain. Nicknamed "honkers" because of their boisterous, honking calls, Canada geese nonetheless announce their presence here from the sky, and they make temporary layovers in both parks during their seasonal migration from the Canadian prairie potholes or the Gulf of Mexico. Look (and listen) for the Canada goose in April and October along the Middle Fork of the Flathead River and at St. Mary Lake in Glacier, and at Lower Waterton Lake in Waterton.

Great Gray Owl
Strix nebulosa

The great gray owl is the largest owl in North America, but it's very rare in Glacier. The national park sits on the southern edge of the great gray's range. A diurnal creature, it's active during the day and is sometimes seen in the late afternoon at the edge of evergreen forests, where it sits and scans open meadows for pocket gophers, snowshoe hares, and other rodents. The great gray has a classic owl shape—an elliptical face, protruding concentric patterns around the eyes, and a puffed and feathered frame. Its plumage, of course, is mottled gray, and the owl's call is the stereotypical *hoo-hoooo*. Males and females are nearly identical in appearance, and it's not uncommon for great grays to forage in pairs. These owls are sometimes seen flying across the highway leading to Many Glacier, and in Waterton along the Akamina Parkway.

Clark's Nutcracker
Nucifraga columbiana

Along many of the highest ridgelines in Glacier, there are contorted trunks of whitebark pine and perhaps a few grizzly bears that owe their existence to Clark's nutcrackers. Every autumn, these big birds with gray, black, and white plumage begin harvesting seeds from whitebark pine cones. The industrious nutcrackers pluck the seeds with their long, sharp bills and store them in the soil for later consumption, but some are always forgotten. Fortunately, these seeds sprout into a new generation of whitebark pine trees that feed grizzly bears, which also rely on the seeds for part of their diet. When you hear Clark's nutcrackers (named after William Clark, co-leader of the famed Lewis and Clark expedition), don't be surprised if they sound like crows, because they are indeed members of the crow family. They have light gray bodies with black wings and a smear of white on the wing tips. Clark's nutcrackers inhabit the alpine environment, near stands of juniper, larch, and whitebark pine. They're often mistaken for gray jays or Canada jays.

Grouse

Most males of the ten North American grouse species engage in ritualistic dances designed to attract the attention of females. In Glacier, there are four non-migratory species that do so. Ruffed grouse (*Bonasa umbellus*) perform what is called "drumming" each spring, flapping their wings wildly while standing on a log. They can be spotted along the Akamina Parkway in Waterton Park just west of the Townsite, and near the U.S.-Canada border on Chief Mountain International Highway (Highway 17). Blue grouse (*Dendragapus obscurus*) generally occupy the high ridge-

lines, where they find an abundance of grasshoppers, junipers, and leafy plants to eat. Male blue grouse will stand on a log and emanate loud "booms" by blowing air in and out of their colorful throat pouches. They can be seen in the remains of the old Garden Wall forest fire burn on Going-to-the-Sun Road. Male spruce grouse (*Canachites canadensis*) fan their tails like peacocks to attract females. Spruce grouse are sometimes visible along the Bowman Lake Road and Camas Road. (See also the chapter on white-tailed ptarmigan.)

Dipper (Water Ouzel)
Cinclus mexicanus

With hundreds of permanent and seasonally charged streams pouring out of the mountains, Glacier offers picture-perfect terrain for dippers, known to many people as "water ouzels." These tiny aquatic birds can skate across flowing water and enthrall the watcher who takes the time to notice their subtle antics and sweet songs. Dippers are creatures of the American West that search for water-borne insects in fast-moving rapids. Slate-gray, they have the frame of a wren and the zeal of a duck. Diving into the chilly water and running along the bottom of a stream—sometimes with their wings extended—dippers have natural oil in their feathers that repels water. The birds are present on most streams, but they're particularly easy to spot upstream from McDonald Falls on upper McDonald Creek.

Five Hikes To Stretch Your Legs and Expand Your Wildlife Watching Adventure

Goat Haunt to Waterton Townsite
Glacier and Waterton Lakes national parks (7 miles, easy to moderate)

One of the most unusual wildlife watching excursions outside your car has an international flavor and involves a boat ride. From Waterton Townsite in Canada, take the scenic commercial boat ride to Glacier's Goat Haunt, located at the southern end of Upper Waterton Lake inside the United States. Remember to have personal identification with you. Once you reach the dock at Goat Haunt, hike approximately seven miles along the west side of the lake back to Waterton Townsite. Bring your binoculars (as well as other necessary provisions). The hike can easily be accomplished in a day if you start early. Along the way, you stand a good chance of seeing bald eagles, osprey, and waterfowl. Near Goat Haunt you may encounter elk, moose and grizzlies. NOTE: On some weekend days, park naturalists lead what is known as the International Peace Park Hike beginning at Waterton Townsite and ending at Goat Haunt, allowing you to take the boat ride back. Inquire with rangers at the information desks in either park.

Iceberg Lake Trail
Glacier National Park (10 miles, moderate)

For those who wish to taste the rugged backcountry terrain that Glacier is known for, and possibly catch a glimpse of two park wildlife symbols—mountain goats and grizzly bears— the ten-mile day hike into Iceberg Lake through the Swiftcurrent Valley is a good trip. The hike begins at the trailhead in a parking lot behind the Swiftcurrent Motel. If you need help finding it, ask a ranger or motel employee. About 2.5 miles up the trail you reach Ptarmigan Falls, which marks the halfway point to Iceberg Lake. Once you get closer to the lake, scan the high slopes for bighorn sheep and mountain goats. NOTE: Be aware that grizzlies often are spotted along this trail so follow the rules of being alert in bear country. The Iceberg Lake Trail is a modest hike in terms of its physical challenge and often attracts families with kids older than eight years old and physically fit seniors who want to spend a day on a hike. The hike is best after the snow clears by the end of June. Near the Many Glacier Hotel, there's also the Swiftcurrent Lake Nature Trail that involves an easy three-mile loop trail around the lake.

Hidden Lake Overlook Trail
Glacier National Park (3 miles, easy to moderate)

Beginning at the Logan Pass Visitor Center along the Going-to-the Sun Road, the Hidden Lake Overlook Trail is short in distance but sweet in the variety of different animals you could see on any given day. Mountain goats are commonly seen along this trail, along with bighorn sheep, marmots, and ground squirrels. Grizzlies and even wolverines—which are extremely rare—have been spotted from the trail. The overlook features a spectacular view of Hidden Lake and the surrounding peaks. It is another three miles down to the lake, dropping some 765 feet from where you begin.

Avalanche Lake Trail/Trail of the Cedars
Glacier National Park (4 miles, easy to moderate)

Just off the northeastern flanks of Lake McDonald along Going-to-the-Sun Road is this hiking path that leads two miles, one way, up the draw of Avalanche Creek to Avalanche Lake. The stroll to the lake is easy and welcoming for wildlife watchers who want modest exercise. One of the best features of this route is that it begins with the Trail of the Cedars, which winds through a forest of towering, thick-trunked hemlocks and cedars, offering the delightfully humid feel of a temperature rainforest. Songbirds are seen along the boardwalk nature trail and as you continue toward Avalanche Lake. Birders should delight in knowing that early in the summer rare breeding Harlequin ducks visit nearby McDonald Creek. Also carefully keep your eyes open for dippers. Be mindful that grizzlies and moose inhabit the area.

Highline Trail
Glacier National Park (7.5 miles one way, moderate)

Among the faithful who adore Glacier over any other national park, the hike along the Highline Trail into Granite Park Chalet is perhaps the quintessential foot excursion. The trailhead is found on Logan Pass along the north side of Going-to-the-Sun Road and covers a one-way distance of more than 7 1/2 miles. Reservations are required long in advance to make an overnight stay at the chalet, but many hard core hikers go in and out in a day. The trail winds its way toward shark-toothed summits and active glaciers, often yielding views of mountain goats, bighorn sheep, wolverine, bears, and birds.

So You'd Like To Know More?

Here's a list of resources that provide valuable information about wildlife watching opportunities in Waterton-Glacier International Peace Park:

Glacier National Park features its own Internet Visitor Center on the World Wide Web at: http://www.nps.gov/glac/home.htm. The site offers all the information you need to plan your vacation in the park. You can also write the park and request a free map. Glacier National Park P.O. Box 128, West Glacier, Montana 59936, phone: (406) 888-7800.

Waterton Lakes National Park also features its own Internet Visitor Center at: http://parkscanada.pch.gc.ca/waterton/. You can also write the park at Waterton Lakes National Park, Waterton Park, Alberta, TOK 2MO, Canada, or phone: (403) 859-2224 (administrative offices) or (403) 859-5133 to reach the visitor center.

The Glacier Natural History Association (GNHA) is a nonprofit cooperating association of the National Park Service. GNHA helps to support Glacier National Park's educational, interpretive, cultural and scientific program needs. Support is generated by sales at bookstores in visitor centers and ranger stations located throughout the Park, in addition to its website bookstore, mail-order catalog, and annual membership program. The association offers a variety of publications and materials, which promote a better understanding of Glacier's diversity of landscapes, animal and plant life, culture, and history. For modest annual dues, members receive a fifteen- percent discount on purchases from GNHA sales outlets, catalog and the website bookstore, accessed at www.glacierassociation.org. Members also receive biannual issues of the association's Goat Notes newsletter, a catalog and invitations to attend annual meetings and other special events. For more information, contact the Glacier Natural History Association via the internet at gnha@glacierassociation.org, or write Glacier Natural History Association, Box 310, West Glacier, MT 59936, or phone: (406) 888-5756.

The Glacier Fund, a subsidiary of the National Park Foundation, was created during the 1990s with a special purpose: To help solicit support for Glacier's protection and upkeep from park lovers in the private sector. Among some of its noble objectives are supporting environmental education, historic building restoration, wildlife research, the historic fleet red buses, trail work, and maintenance of the park's world famous backcountry chalets. By supporting both the Glacier Natural History Association and the Glacier Fund, individuals and businesses can make a profound difference in ensuring that Glacier endures for future generations to enjoy. For more information about the Glacier Fund, write: The Glacier Fund, Glacier National Park, West Glacier, MT 59936, or phone: (406) 888-7910. The fund's web address is: http://www.nps.gov/glac/partners/glacfund.htm

The Glacier Institute is a private, nonprofit outdoor education organization dedicated to connecting children and adults with the natural and cultural wonders of the Crown of the Continent Ecosystem, the heart of which is Glacier National Park. Since 1983, the institute has been providing hands-on, field-based educational adventures—including classes that focus on wildlife— in the national park for people who come from all around the world. You can first visit the Glacier Institute's web site at: http://www.glacierinstitute.org/. Or write: Glacier Institute, 137 Main St., Kalispell, Montana 59904, or phone: (406) 755-1211.

The Waterton Natural History Association is a cooperating association dedicated to supporting and enhancing the values and purposes of Waterton Lakes National Park through personal service, publications and educational programs. For more information on the program listings, write Waterton Natural History Association, Box 145, Waterton Lakes, Alberta, T0K 2M0, Canada.